My
iPad
for Seniors

EIGHTH EDITION

Que®

Michael Miller

My iPad® for Seniors, Eighth Edition

Copyright © 2021 by Pearson Education, Inc.

ISBN-13: 978-0-13-682429-9

ISBN-10: 0-13-682429-3

Library of Congress Control Number: 2020946308

1 2020

Trademarks

AARP is a registered trademark.

Warning and Disclaimer

Special Sales

For information about buying this title in bulk quantities, or for special sales opportunities (which may include electronic versions; custom cover designs; and content particular to your business, training goals, marketing focus, or branding interests), please contact our corporate sales department at corpsales@pearsoned.com or (800) 382-3419.

For government sales inquiries, please contact governmentsales@pearsoned.com.

For questions about sales outside the U.S., please contact intlcs@pearsoned.com.

Editor-in-Chief
Brett Bartow

Executive Editor
Laura Norman

Associate Editor
Chhavi Vig

Marketing
Stephane Nakib

Director, AARP Books
Jodi Lipson

Editorial Services
The Wordsmithery LLC

Managing Editor
Sandra Schroeder

Senior Project Editor
Lori Lyons

Copy Editor
Charlotte Kughen

Indexer
Ken Johnson

Proofreader
The Wordsmithery LLC

Technical Editor
Jeri Usbay

Editorial Assistant
Cindy J. Teeters

Cover Designer
Chuti Prasertsith

Compositor
Bronkella Publishing

Graphics
T J Graham Art

Contents at a Glance

Table of Contents

4 Making Your iPad More Accessible   **79**

7   Controlling Your iPad—and More—with Siri   **133**

8   Installing and Using Apps   **149**

About the Author

Michael Miller is the popular and prolific writer of more than 200 non-fiction books, known for his ability to explain complex topics to everyday readers. He writes about a variety of topics, including technology, business, and music. His best-selling books for Que include *My TV for Seniors*, *My Facebook for Seniors*, *My Social Media for Seniors*, *My Internet for Seniors*, *My Smart Home for Seniors*, *My Windows 10 Computer for Seniors*, *Easy Computer Basics*, and *Computer Basics: Absolute Beginner's Guide*. Worldwide, his books have sold more than 1.5 million copies.

Find out more at the author's website: www.millerwriter.com

Follow the author on Twitter: molehillgroup

Dedication

To my musician friends and family here in Minnesota who make it fun to play.

Acknowledgments

Thanks to all the folks at Que and Pearson who helped turn this manuscript into a book, including Laura Norman, Chhavi Vig, Charlotte Kughen, Lori Lyons, and technical editor Jeri Usbay. Thanks also to Jodi Lipson and the kind folks at AARP for adding even more to the project.

About AARP

AARP is a nonprofit, nonpartisan organization, with a membership of nearly 38 million, that helps people turn their goals and dreams into *real possibilities*™, strengthens communities, and fights for the issues that matter most to families, such as healthcare, employment and income security, retirement planning, affordable utilities, and protection from financial abuse. Learn more at aarp.org.

Note

Most of the individuals pictured throughout this book are the author himself, as well as friends and relatives (used with permission) and sometimes pets. Some names and personal information are fictitious.

We Want to Hear from You!

As the reader of this book, *you* are our most important critic and commentator. We value your opinion and want to know what we're doing right, what we could do better, what areas you'd like to see us publish in, and any other words of wisdom you're willing to pass our way.

We welcome your comments. You can email to let us know what you did or didn't like about this book—as well as what we can do to make our books better.

Please note that we cannot help you with technical problems related to the topic of this book.

When you email, please be sure to include this book's title and author, as well as your name and email address. We will carefully review your comments and share them with the author and editors who worked on the book.

Email: community@informit.com

Reader Services

Register your copy of *My iPad for Seniors* at informit.com/register for convenient access to downloads, updates, and corrections as they become available. To start the registration process, go to informit.com/register and log in or create an account.* Enter the product ISBN, 9780136824299, and click Submit.

*Be sure to check the box that you would like to hear from us in order to receive exclusive discounts on future editions of this product.

Figure Credits

Cover	©-strizh- | Shutterstock.com Product images courtesy of Chuti Prasertsith
Apple apps screenshots	© Apple Inc.
AARP webpage screenshots	Courtesy of AARP
Google Chrome screenshots	Google and Google logo are registered trademarks of Google LLC
Safari browser screenshots	Safari Browser, Apple Inc.
Facebook screenshots	Courtesy of Facebook Inc.
LinkedIn screenshots	LinkedIn Corporation © 2019
Pinterest screenshots	© Pinterest
Twitter screenshots	© Twitter Inc.
Pandora screenshots	© 2019 Pandora Media, Inc.
Spotify screenshots	© 2018 Spotify AB
Amazon screenshots	© 1996–2019, Amazon.com, Inc.
Hulu screenshots	© 2019 Hulu
Netflix screenshots	© Netflix Inc.
YouTube screenshots	© 2019 YouTube
CBS All Access	© 2020 CBS Interactive
Disney+	© Disney. All Rights Reserved.
HBO Max	©2020 WarnerMedia Direct, LLC. All Rights Reserved. HBO Max™ is used under license.
Microsoft Word app	Microsoft® and Windows® are registered trademarks of the Microsoft Corporation in the U.S.A. and other countries. Screenshots and icons reprinted with permission from the Microsoft Corporation.
TripAdvisor	© 2020 TripAdvisor LLC All rights reserved.
Yelp	Copyright © 2004–2020 Yelp Inc.
Daily Yoga	© 2012–2020 Daily Yoga Culture Technology Co. Ltd
MyFitnessPal's Calorie Counter & Diet Tracker	© 2020 Under Armour, Inc. All Rights Reserved.

MyPlate Calorie Counter	© 2019 Livestrong.com
Allrecipes Dinner Spinner	© 2015 Allrecipes.com
Yummly Recipes and Shopping List	© 2020 Yummly Inc.
Drugs.com Medication Guide	© 2016 Drugs.com
Medical Dictionary by Farlex	© 2019 Farlex, Inc.
WebMD	© 2020 WebMD, LLC
Apple Arcade games	© Apple Inc.

In this chapter, you learn about the various iPad models and how to unbox and set up your new iPad.

→ Choosing the Right iPad for You

→ Unboxing Your New iPad

→ Powering Up Your iPad for the First Time

→ Accessorizing Your iPad

Buying and Unboxing Your iPad

As you probably know, Apple's iPad is a *tablet computer*—that is, a small computer in the shape of a handheld tablet. Instead of a keyboard and separate display screen, you have a touch-sensitive display. Not only do you view whatever is on the screen, you also can touch or tap or otherwise poke and prod the screen with your fingertips to make things happen. Instead of tapping the keys on a physical keyboard, you tap the virtual keys and buttons on the display, and the tablet reacts.

You can use an iPad to browse and search the Web, keep in touch with people through Facebook and other social media, send and receive email and instant messages, participate in video chats, watch movies and TV shows, listen to music, take and edit photographs and videos, play games, and control smart home devices. You can even use an iPad for word processing, creating spreadsheets, and managing other productive activities.

For many people, an iPad is a viable replacement for a notebook or desktop computer because an iPad can do so many things. Or you may view a tablet as a useful supplement to your smartphone, one that's just easier and more comfortable to use for many tasks.

Choosing the Right iPad for You

Apple has offered various models of iPads since its introductory model a decade ago, with new types and sizes being released annually.

All of Apple's iPads share a number of common features. They all have touch-screens, so you can control what you see with a tap or drag of your finger. They all have built-in speakers for listening, and most have a jack to which you can connect your earphones or headphones. They all have cameras in the rear for shooting normal photos and in the front for shooting selfies or doing video chats. They all have built-in storage to store those photos you take as well as other data and apps. All will last about 10 hours on a battery charge, depending on your usage. And they all have built-in Wi-Fi, so you can wirelessly connect to the Internet.

Beyond that, the models differ mostly in terms of screen size and resolution, stor-age capacity, processing speed, and a few extra features, such as cellular connec-tivity (so you can connect your iPad to your mobile phone network). Naturally, the bigger models with more speed, storage, and features cost more than the smaller, less fully featured ones.

So which model you choose depends on a number of factors—what size screen you want, how much storage capacity you need, what you want to do with the iPad, and how much money you want to spend. Let's look at the current models available, as of Fall 2020.

iPad mini

The smallest (but not the least expensive) iPad in Apple's lineup is the iPad mini. The current fifth-generation mini is smaller than the original iPad, with a 7.9-inch screen (compared to the traditional iPad's 9.8-inch screen). This makes the mini both smaller (just 8 inches high by 5.3 inches wide) and lighter (.66 pound) than the traditional iPad, which a lot of people prefer; you don't need to use both hands to hold it.

The fifth-generation iPad mini

Of course, the smaller display also makes it a little more difficult to read what's on the screen. The text on some web pages gets a tad small for some readers who might be more comfortable with the larger screen on the traditional iPad. That said, the screen on the fifth-generation iPad mini has a higher resolution than the regular-sized model (measured in terms of pixels, or individual screen elements), which provides a sharper picture.

Retina Display

All of Apple's latest iPad models feature what the company calls the *Retina display*. The Retina display has more pixels per square inch, which makes text and images appear extremely crisp onscreen. In general, the higher the resolution (the more pixels per inch), the sharper the display.

Price-wise, the iPad mini is actually more expensive than the larger iPad model (which I discuss next). Apple currently offers two Wi-Fi iPad mini models, one with 64GB of storage for $399, and one with 256GB of storage for $549. Both are compatible with the (optional) Apple Pencil stylus device.

Cellular Connectivity

All iPads come with Wi-Fi built in, so you can connect to the Internet from your home Wi-Fi network or any public Wi-Fi hotspot. Select models are also available that add cellular connectivity, so you can connect the iPad to your mobile phone's data network when Wi-Fi is not available. Cellular connectivity adds about $130 to the price of a comparable Wi-Fi-only model; because of this price differential, the noncellular iPads are Apple's best sellers.

iPad

Apple's flagship tablet is simply called the iPad. This model, Apple's most popular, comes with a large 10.2-inch Retina display, which makes it ideal for those of us with aging eyes. It's 9.8 inches high × 6.8 inches wide and weighs 1.08 pounds.

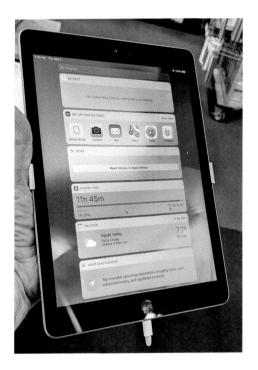

Apple's eighth-generation iPad

The new eighth-generation version of the iPad includes support for the (optional) Apple Pencil stylus and Smart Keyboard. It comes in 32GB and 128GB versions and is priced at just $329 and $429, respectively.

iPad Air

The iPad Air is a tad faster, has a display that's a tad larger, and is a tad more versatile than the regular iPad. The fourth-generation iPad Air has slightly larger exterior dimensions than the regular iPad—9.74 inches high × 7 inches wide—but it weighs a bit less at an even 1.0 pound. It has a 10.9-inch Retina display and 2360 × 1640 pixel resolution. (That calculates to the same 264 ppi as the slightly smaller iPad, however.) The iPad Air also features a slightly faster processor than the regular iPad.

Like the iPad Pro, the iPad Air does away with the Home button, instead relying on touch gestures. For security, it features Touch ID built into the Top button. The fourth-generation iPad Air is also compatible with the (optional) Apple Pencil, Smart Keyboard Folio, and Magic Keyboard.

This iPad Air comes in two versions—with 64GB and 256GB storage. They're priced at $599 and $749, respectively.

iPad Pro

Apple's top-of-the-line iPad is the iPad Pro. This is a fully featured model, available in two screen sizes, that is optimized for business productivity. When coupled with the Apple Pencil and Smart Keyboard Folio or Magic Keyboard, all optional, the iPad Pro can function as a viable replacement for a traditional desktop or laptop PC.

The iPad Pro—no Home button

Both iPad Pro models also have additional features not found on lower-priced iPads. They do away with the traditional Home button, instead relying on touch gestures and Face ID technology. They both feature USB-C connectors instead of the older Lightning connectors found on the non-Pro models. Screen resolution is 264 ppi for both models, and they both offer True Tone display technology and two cameras in the rear—a 12-megapixel camera with f1.8 wide-angle lens and a 10-megapixel camera with a f2.4 ultra-wide angle lens.

Because of the enhanced productivity (and slightly larger screens), iPad Pro models aren't cheap. The 11-inch iPad Pro (slightly larger than both the iPad and iPad Air) starts at $799 for 128GB of storage, and goes up to $1,299 for a 1TB model. The larger 12.9-inch model, which is even better suited for office use, starts at $999 for 128GB storage and goes up to $1,499 for 1TB storage.

>>>Go Further
iPadOS—THE iPAD OPERATING SYSTEM

All iPad models are powered by a special operating system designed just for mobile devices, dubbed iPadOS. It's similar to the iOS operating system on Apple's iPhones, but it's further tweaked for tablet use.

The current version of iPadOS, introduced in September 2020, is iPadOS 14. Prior to iPadOS 13, iPads had run the standard iOS operating system; the last version of iOS for iPads was iOS 12 (which explains why the first version of iPadOS after iOS 12 was iPadOS 13).

All new iPads come with iPadOS 14 preinstalled; if you have an older model, you can easily upgrade to iPadOS 14 by opening the Settings app on your iPad, tapping General, and then tapping Software Update.

This book covers iPads running iPadOS 14. If your iPad is running an older version of iOS, you can either upgrade your device or purchase a previous version of this book.

Which iPad Should You Buy?

As noted previously, which iPad is best for you depends on how you intend to use it, along with what size screen you want and how much you want to spend.

In general, the most popular model with both home and casual users is the 10.2-inch iPad. At $329 for the base model, it's the lowest price of all the iPad models.

Now, some home and casual users prefer the iPad mini simply because of the size. Given that it's priced higher than the larger base model iPad, however, it's become a niche model aimed at smaller children and schools.

If you want a slightly larger screen than the standard iPad, along with slightly faster performance, the iPad Air is a good choice. It costs almost twice as much as the standard iPad, however, so there's that.

If you want to do serious work on your tablet or are using your tablet in your business as a replacement for a notebook or desktop computer, then you should consider the iPad Pro models. Know, however, that you'll spend more (a lot more) for the larger screens and enhanced functionality. Nice as they are, the iPad Pros are just too pricey for many home and casual users.

All current model iPads run the same iPadOS 14 operating system, and all run pretty much the same apps you can find in Apple's App Store. The choice really comes down to a matter of size, storage capacity, and what you want to use it for; choose the one that best fits your needs.

Unboxing Your New iPad

Whichever model iPad you choose, it comes in a stylish white box with minimal text and graphics. (Apple is all about the style!) What's inside that box? Let's take a peek!

What's Inside the Box

All iPad models come in similar packaging. Open the box and you see the iPad itself. Take out the iPad and remove the plastic protector sheets.

Unboxing a brand-new iPad Pro

Charging Your iPad

Your new iPad's battery probably has a decent charge out of the box, so you can use it right away. However, you'll soon want to recharge the battery to get maximum use out of your new device. Learn how in Chapter 2, "Getting Started with Your iPad."

Underneath the iPad are a white cable and power adapter, which you use to both charge your iPad and connect it to a computer. The iPad, iPad Air, and iPad mini models feature a Lightning-to-USB cable; the iPad Pro models feature a USB-C-to-USB cable. The power adapter, of course, plugs right into any wall outlet.

Powering Up Your iPad for the First Time

Now is the moment of truth—time to power up your iPad for the very first time! It takes about 5 to 10 minutes to get everything set up, but it's as simple as following the instructions you see onscreen.

During the course of the initial setup, you're asked to supply some or all of the following information:

- Your language and country or region. This determines how specific information, such as date and time, appears on your device.
- The name and password for your home wireless network or nearby Wi-Fi hotspot for your iPad to connect to the Internet.
- On those iPads with a Home button, you need to register your fingerprint to enable Touch ID fingerprint log in.
- On those iPads without a home button, you need to register your face (using the iPad's built-in camera) to enable the Face ID facial recognition log in.
- On all iPads, you also need to create a six-digit passcode for your device.

Passwords and PINs

Learn more about protecting your iPad with passwords, PINs, Touch ID, and Face ID in Chapter 6, "Keeping Your iPad Safe and Secure."

- If you have another Apple device handy (and it's running iPadOS 13, iOS 13, iOS 12, or iOS 11), you're prompted to use the Automatic Setup features to wirelessly transfer settings from the other device to your iPad. Just hold the two devices close together until they're properly "paired," and then let Automatic Setup do all the work for you.

- If you have an Apple ID for another Apple device (another iPad, iPhone, or Mac), you can use it to log in and transfer some system settings. You also can go this route if you have an older iPad from which you want to transfer settings and apps.

- If this is your first Apple device, you're prompted to create a new Apple ID. This is free.

You're also prompted to set up Apple Pay (for using your iPad for hands-free and online payments), iCloud Keychain (to save usernames and passwords), and Siri (for voice commands). All of these are optional, and you can configure them during initial setup or later if you don't have time to do it right away. You'll also have to accept Apple's Terms and Conditions; you have to do this before Apple will let you use your new iPad.

Turn It On—for the First Time

Turning on your new iPad is a simple process—as easy as pressing a button.

(1) Turn on your iPad by pressing and holding the Top button (at the top of the case) until you see the Apple symbol on the screen.

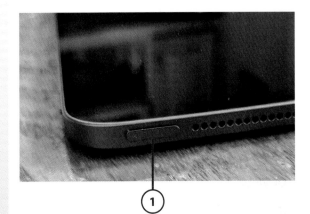

(2) You see the word "Hello" in many languages. Swipe up from the bottom of the screen or press the Home button (at the bottom front of the case), if your iPad has one, to continue. You should now follow the onscreen instructions to complete the initial setup.

Accessorizing Your iPad

Apple and a number of third-party companies make a variety of accessories you can add to your iPad. These accessories protect your iPad and make it more functional and more personalized.

Perhaps the most popular iPad accessory is an iPad case. A rugged case is essential if you want to protect your iPad against accidental bumps and drops, which can easily damage your expensive device. One particularly useful type of case is the *smart cover*, which automatically puts your iPad to sleep when closed and wakes it up when opened. You can find smart covers from Apple and a variety of third parties, many of which feature foldable backs to function as stands for your device.

If you want to use your iPad for office work, you may need something other than the onscreen keyboard for typing. If you're the proud owner of an iPad Pro, consider investing in Apple's Smart Keyboard Folio. Functioning as a keyboard, cover, and folding stand, this keyboard connects to your iPad Pro via the device's wireless Smart Connector. It sells for $179 for the 11-inch iPad Pro, or $199 for the 12.9-inch model.

Apple's Smart Keyboard, connected to an iPad Pro

Apple's Magic Keyboard goes one step further and includes a trackpad and back-lit keys that turn your iPad Pro into a full-fledged desktop computer. It raises the iPad off the desk into a position much like that of a flatscreen computer monitor. It sells for $299 for the 11-inch model and $349 for the 12.9-inch model.

Apple's Magic Keyboard, complete with trackpad

If you have a regular (non-Pro) iPad, there are lots of other wireless keyboards available—priced substantially lower, unsurprisingly. Some of these are free-standing keyboards; others come as part of a keyboard/cover/stand. Prices range from around $25 up to $100 or more.

Another useful accessory for iPad owners is the Apple Pencil, a wireless stylus that resembles a physical pencil and enables you to write or draw on the screen. You also can use it as a stylus to select items onscreen. The first-generation Apple Pencil, which works on most current non-Pro iPads, runs $99. The second-generation Apple Pencil, which works with the newest iPad Pro models, runs $129.

Using an Apple Pencil with an iPad Pro

Using the Apple Pencil, Smart Keyboard, and Magic Keyboard

Learn more about using the Apple Pencil, Smart Keyboard Folio, and Magic Keyboard in Chapter 22, "Using Pencils, Keyboards, and Trackpads."

Alternatively, you can opt for a third-party stylus that connects to any iPad via Bluetooth. This type of stylus works just about the same and costs a lot less than the Apple Pencil.

If you like to listen to music or watch videos without disturbing others nearby, invest in wireless headphones or earphones. Or if you want to use your iPad to host the music for your entire living room, get one or more wireless Bluetooth speakers and turn your iPad into the centerpiece of an honest-to-goodness home audio system.

Where can you find all these accessories? You can find iPad accessories from a variety of different manufacturers at Best Buy, Target, Walmart, and other traditional retailers, as well as Amazon.com and other online retailers. Apple also sells a bevy of iPad accessories (both its own and third-party items) on its website at www.apple.com/shop/ipad/ipad-accessories. Prices vary from store to store, of course, so make sure you shop around for the best price!

2

Getting Started with Your iPad

Now that you've taken your new iPad out of the box, charged it up, and performed the initial setup, it's time to start using it. To do so, you need to know what's what and what's where on the iPad itself—and what you need to do to operate the darned thing!

Getting to Know Your iPad

Your iPad is a large, flat tablet with a screen on one side and a fairly plain back on the other. You can hold it either vertically (portrait mode) or horizontally (landscape mode); the screen flips to accommodate how you're holding it.

There are physical buttons along the top and sides of the iPad and on the bottom front. Let's look at these buttons now.

Front

Looking at the iPad from the front, the first thing you see is the screen. This is the touchscreen (technically, Apple calls it a Multi-Touch Display), which is how you perform most of the device's operations—by literally touching the screen with your fingers. I go over all the multi-touch gestures (tapping, swiping, pinching, and more) in the "Learn Essential Multi-Touch Gestures" section later in this chapter.

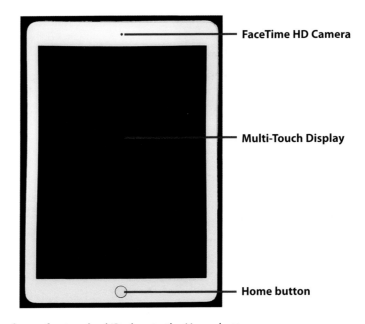

The front of a standard iPad; note the Home button

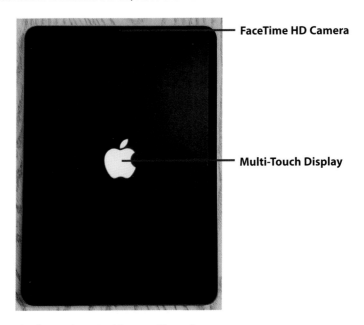

The front of an iPad Pro; no Home button

Some iPads have a round button just beneath the screen; this is called the Home button. You press this button to return at any time to the iPad's Home screen. The Home button also doubles as the Touch ID sensor; if you configure your iPad to unlock with your fingerprint, this is where you press your finger to unlock the device.

Touch ID and Face ID

Learn more about fingerprint and face unlocking in Chapter 6, "Keeping Your iPad Safe and Secure."

Not all iPads have a Home button. The iPad Air and all iPad Pro models have a clean front with no Home button. The fourth-generation iPad Air uses Touch ID built into the Top button to unlock the device; all iPad Pro models use Face ID to unlock. All models without a Home button use touch gestures for navigation. For example, to go to the Home screen, swipe up from the bottom of the screen until the Home screen appears. (Again, I discuss these gestures later in this chapter.)

Directly above the screen is a small hole. This is the lens for the iPad's front-facing camera—the one you use to take selfies or conduct video chats. Apple calls this the FaceTime HD camera, and you probably don't want to cover it up.

Top

There are several items along the top edge of your iPad.

Some models, like the iPad Pros, have two speakers on the top of the device; others don't. One or more microphones are near the center of the top of the unit.

In addition, all iPads include a small button for turning your iPad on or off on the top right of the device; this is called either the Top or the On/Off button. You press this button to wake your iPad when it's asleep or put it to sleep if you're using it. On the iPad Air, Touch ID is built into the Top button.

The top of an iPad Pro

Some iPads have a 3.5mm headset jack for connecting earphones or headphones. Newer models, including the iPad Air and all iPad Pro models, do away with this traditional headset jack. On these models, you need headphones or earphones that connect via USB-C at the bottom of the iPad, or a set of wireless Bluetooth earphones.

Sides

There's nothing on the left side of your iPad. On the right side, however, you find the up and down volume buttons. Use these controls to raise and lower the iPad's volume.

Volume up

Volume down

The right side of an iPad Pro

Cellular Models

If your iPad offers cellular connectivity (in addition to the normal Wi-Fi wireless), it has a removable tray on the right side of the unit. This tray contains the iPad's Nano-SIM card; insert the included removal tool (or a plain old paperclip) into the small hole to eject the tray.

Bottom

There are several important items along the bottom edge of your iPad, too.

In the very middle bottom is a small connector. On some models this is a Lightning connector; on others, it's a USB-C connector. They look similar and do similar things.

USB-C
connector

Speakers

The bottom of an iPad Pro

You use the Lightning or USB-C connector to connect the cable that came with your iPad. The other end of the cable connects to the power adapter that you plug into the wall. (You can also plug it into your computer to transfer files.)

There are two speakers on the bottom of the iPad on either side of the Lightning or USB-C connector. Don't cover up these speakers or you'll adversely affect the sounds you hear.

Back

Finally, turn your iPad over and examine the back of the unit. The most obvious thing on the back is the big Apple logo, but that really doesn't do anything.

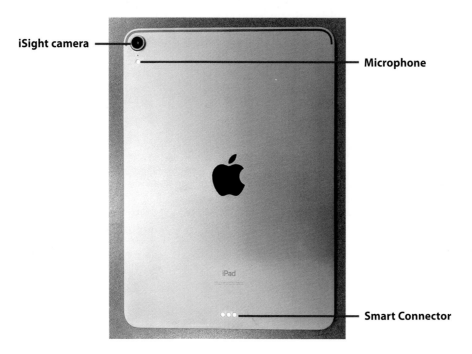

iSight camera

Microphone

iPad

Smart Connector

The back of an iPad Pro

Instead, direct your attention to the top-left corner. The big round thing you see there is the main camera. Apple calls this the iSight camera, and you use it to take pictures and videos of others.

Some iPad Pro models have more than one camera, including one for ultra-wide angle shots. Multi-camera models have a square module in the top-left corner that houses the cameras.

Just beneath the camera(s) on some models is a small hole that contains another microphone. On other models, this may be in the top middle. Don't cover up the microphone—or the camera, for that matter.

Finally, on iPad Pro models, you see three round metallic dots at the very bottom of the back. These are the magnetic connectors for your iPad's Smart Connector. These magnetically connect so-called "smart" accessories, such as the iPad Pro Smart Keyboard, to your iPad Pro.

Turning Your iPad On and Off

Turning your iPad on and off is a little more complex than you might think. That's because you can completely power off your unit or just put it to sleep. When the unit is completely powered off, it takes a minute or so to power back on. When it's in sleep mode (or "locked," as Apple puts it), it can immediately come back to life with the same apps running as when you put it to sleep.

Power On Your iPad

When your iPad is completely powered off (as it is when you first remove it from its box), you need to power it back on.

① Press and hold the On/Off button until the Apple logo appears onscreen.

(2) What happens after the Apple logo appears depends on your particular model of iPad and how you've opted to unlock the device. (Not shown.) See the "Unlock Your iPad" tasks, next, to proceed and display the Home screen.

Setting Up Unlock Methods

Learn how to set up the different ways to unlock your iPad in Chapter 6.

Unlock Your iPad with Face ID

To reawaken a sleeping iPad, all you have to do is unlock it. If you have an iPad Pro, you can unlock your device with Face ID, which uses facial recognition technology. Obviously, you need to configure this first (which you probably did when you first set up your iPad), but then it's a snap to use.

(1) Tap the iPad's screen to wake it.

(2) Stare at the screen until the lock icon changes from closed to open.

(3) Swipe up from the bottom of the screen to unlock.

Unlock Your iPad with Touch ID

If your iPad has a Home button, you can use Touch ID to unlock your phone via fingerprint. (Again, this needs to be set up in advance.)

(**1**) Tap the Home button.

(**2**) Rest your finger or thumb (which-ever you've registered) on the Home button until the iPad unlocks.

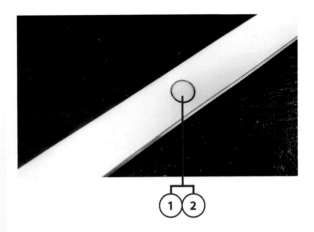

Unlock Your iPad with a Passcode

If you've configured your iPad with a passcode, you can unlock your iPad that way, too.

(**1**) Tap the Home button or, on models without a Home button, tap the screen.

(**2**) Enter your passcode.

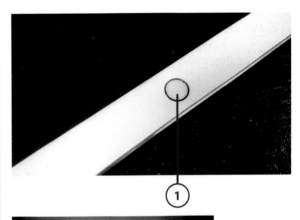

Wake Your iPad with No Lock

Although it's not recommended, you don't have to set up a passcode, Touch ID, or Face ID to unlock your iPad. This leaves your iPad entirely unsecured so that anyone can use it. If you haven't set up any lock protection, all you have to do to wake it is tap the Home button or, on models without a Home button, tap the screen.

Unlock an iPad Air

If you have a fourth-generation iPad Air, it doesn't have a Home button and doesn't use Face ID. Instead, Touch ID is built into the Top button; press your finger to the Top button to unlock.

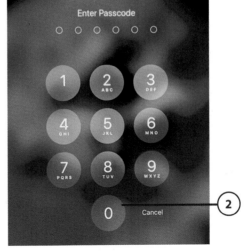

Lock Your iPad

You put your iPad to sleep (but not turn it completely off) by locking the device. This is how you most often "turn off" your iPad.

(1) Press the On/Off button.

(2) The screen goes blank, and the iPad is now locked (not shown).

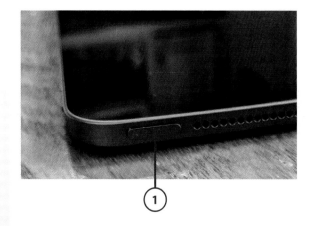

Inactivity Lock

By default, your iPad locks automatically if you haven't touched the screen for two minutes. You can change this auto-lock time on the Settings screen, as discussed in Chapter 3, "Personalizing the Way Your iPad Looks and Works."

Power Off Your iPad

Turning off the iPad shuts down the entire device. You have to power it back up again the next time you want to use it.

(1) On a regular iPad, press and hold the On/Off button for a few seconds; on an iPad Pro, press and hold the On/Off button and either the Volume Up or Volume Down button for a few seconds. Keep holding until the slider appears onscreen.

(2) Drag the Slide to Power Off slider to the right. The iPad powers down.

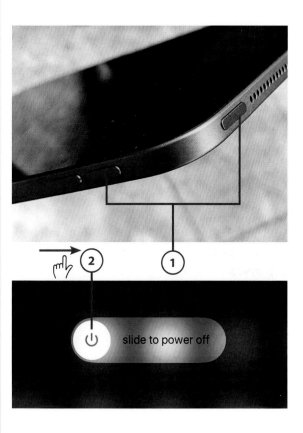

Force a Restart on a Regular iPad

On rare occasions, an iPad may become frozen—tapping the screen does nothing. In this instance, you need to force a restart of your device. Here's how you do it on an iPad with a Home button:

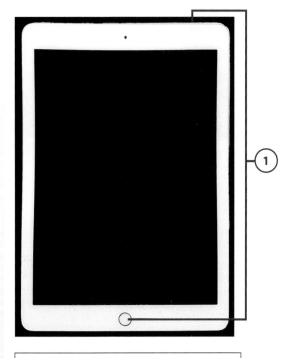

(1) Press and hold both the Top and Home buttons at the same time for at least 10 seconds.

(2) When the Apple logo appears onscreen, your iPad is restarting.

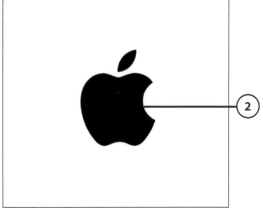

Force a Restart on an iPad Pro

On iPad models, such as the iPad Pro, that don't have a Home button, you can't use the previous method to force a restart. Instead, follow these steps:

1. Press and release the Volume Up button.

2. Press and release the Volume Down button.

3. Press and hold the On/Off button until you see the Apple logo appear onscreen. This means your iPad is restarting.

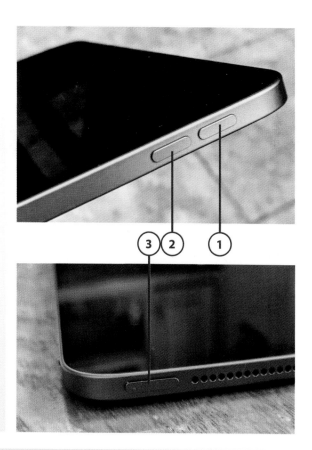

>>>Go Further
SMART COVERS

Many companies (Apple included) sell so-called "smart" covers and cases for the various models of iPads. A smart cover or case not only protects your device, but interacts with your iPad to wake it when you open the cover and put the iPad to sleep when you close the cover.

Using Your iPad

Okay, now you can turn your iPad on and off (and lock it and unlock it, too). But what do you do with it once you turn it on?

Learn Essential Multi-Touch Gestures

You operate your iPad with your finger(s). You can perform different operations with different gestures.

Most everything you can do onscreen is based on a half dozen basic gestures. These are detailed in the following table.

Basic Touch Gestures

Gesture	Looks Like	Description
Tap		Touch and quickly release a point on the screen with your finger.
Press		Touch your finger to the screen and hold it there.
Drag		Touch and hold your finger on an item on the screen, then slowly move your finger to drag the item to a new position. Release your finger to "drop" the item to the new position.
Swipe		This is like dragging, except faster. Place the tip of your finger on the screen, then quickly move it in the appropriate direction. You use swiping to scroll up or down a screen or move the screen left or right.
Pinch		Position your thumb and forefinger apart on the screen and then pinch them together. This is typically used to zoom out of a page or image.
Expand		This is the opposite of pinching. Position your thumb and forefinger together on the screen and then slide them apart. This is typically used to zoom into a page or an image.

Perform Common Operations

You use variations of these basic gestures to perform common operations on your iPad. Here are some of the most useful:

- Return to the Home screen by swiping up from the bottom of the screen until the Home screen displays. (You can also pinch with all five fingers on the screen.)

- Display the Cover Sheet (which is similar to the Lock screen, but the iPad isn't really locked) by pressing the top middle of the screen and pulling downward.

- Display the Search pane by swiping down in the middle of the screen (but not from the top middle!).

- Display the iPad's Today View by swiping from left to right on any Home screen. This screen displays headlines, alerts, and notices from various apps. (Today View may be locked into place when in landscape mode, as described in Chapter 3.)

- Display the Control Center by swiping from the top-right corner of the screen toward the middle. (You use the Control Center to adjust basic device settings.)

- Display all open apps (called the App Switcher) by swiping up from the bottom edge of the screen and then to the right without lifting your finger as you swipe. (On iPads with a Home button, you also can display the App Switcher by tapping the Home button twice.)

- Switch from one open app to another by swiping left to right (or right to left) with four or five fingers.

- With any app open, swipe up slightly from the bottom of the screen to display the Dock.

- Open a second app in split screen mode by displaying the Dock and then dragging the icon for that app up and to the side of the screen.

- While you're in split screen mode, tap and drag a file, photo, or selected piece of text from one app to another.

- Within an app, return to the top of a long page by tapping the top menu bar.

- Zoom in and out of a document (or some web pages) by double-tapping the page.

In addition, the following gestures work when you're editing a document:

- Display a formatting bar by tapping and holding the screen with three fingers.

- Select a word by double-tapping it.

- Select a sentence by triple-tapping it.

- Select a paragraph by tapping it four times.

- Split the onscreen keyboard into two halves by expanding the keyboard with your thumb and forefinger. (Some people find this makes typing a little easier.) Pinch your fingers back together to rejoin the two halves into a whole keyboard.

Navigate the Home Screen

On your iPad, the Home screen is where everything starts. When you press the Home button (if your iPad has one) or swipe up from the bottom of any page, you see the Home screen—or, more accurately, the first of several Home screens. All Home screens display a grid of icons that represent the apps installed on your device, along with a Dock of your most recent apps at the bottom of the screen. You can have as many Home screens as you need to display all your apps.

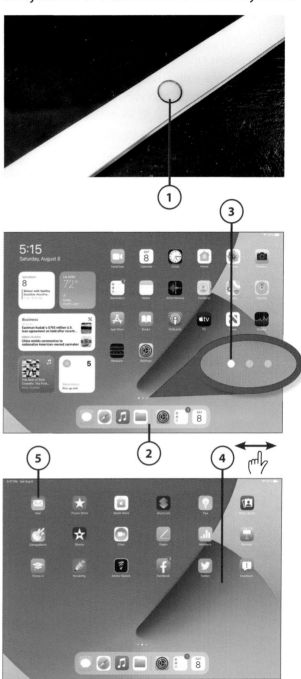

1. Press the Home button or swipe up from the bottom of any screen to display the first Home screen.

2. At the bottom of the screen is the Dock, where app icons can be "docked" so that they appear on every Home page.

3. The number of Home screens on your iPad is indicated by a grouping of dots above the Dock; you see as many dots as you have Home screens. The screen you're currently viewing is the solid white dot in the group.

4. Swipe left or right to view the next or previous Home screen.

5. Tap an icon to open the corresponding app.

View Notifications on the Cover Sheet

As you recall, the iPad's Lock screen can display system notifications and messages from selected apps. These notifications are collectively called the Cover Sheet, and you can view them at any time without having to unlock your device.

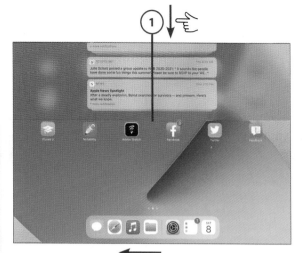

1. Press and pull down from the middle top of any Home screen to display the Cover Sheet. (Or, if your iPad is locked, swipe up on the screen.)

2. Tap any notification to open the notification and accompanying app. *Or…*

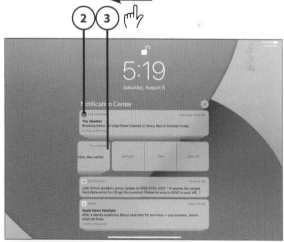

3. Swipe left on any notification to bring up options to view, manage, or clear that item.

4. Swipe up on the Cover Sheet to return to the previously viewed screen.

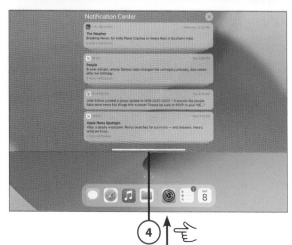

Manage Open Apps

You open apps on your iPad by tapping the app's icon on one of the Home screens or in the Dock. You can have multiple apps open at the same time and easily switch between them.

1. To view all open apps, press the Home button twice if your iPad has a Home button. *Or...*

2. From the bottom of any screen, swipe up, then to the right, then up again without lifting your finger as you swipe.

3. You now see all your open apps. (The open apps are collectively called the App Switcher.) Tap an app to go to that app.

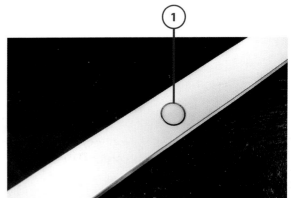

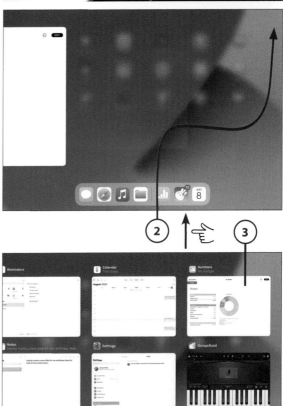

(4) Drag an app up and off the
screen to close it.

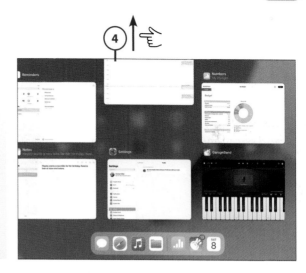

Understand the Status Bar

Running across the top of every iPad screen is a transparent status bar. This status bar displays icons for various system operations and provides information that shows you how your iPad is working.

The following table details the more common icons you'll find on the status bar and what they mean.

Status Bar Icons

Icon	Name	Description
🔒	Lock	The device is locked.
🔋	Battery	Shows the battery level or charging status.
⏰	Alarm	An alarm is set.
☀	Activity	There is currently app or network activity.
🔄	Orientation Lock	Screen orientation is locked.
📶	Wi-Fi	The iPad is connected to a Wi-Fi network; the more bars, the stronger the connection.

Icon	Name	Description
✈	Airplane Mode	Airplane Mode is engaged (Wi-Fi and Bluetooth are both turned off).
☾	Do Not Disturb	Do Not Disturb mode is turned on.
◎	Personal Hotspot	A cellular-equipped iPad is providing a personal hotspot for other devices.
↻	Syncing	The iPad is connected to your computer and syncing with iTunes.
VPN	VPN	The iPad is connected to a virtual private network (VPN).
◁	Location Services	An app is using Location Services to establish the current location.
✶	Bluetooth	If the icon is blue or white, Bluetooth is turned on and the iPad is paired with a device, such as a Bluetooth headset. If the icon is gray, Bluetooth is turned on and the iPad is paired with a device, but the device is turned off or out of range.
▯	Bluetooth Battery	Shows the battery level of the connected Bluetooth device.

Cellular Icons

If you have an iPad with cellular connectivity (Wi-Fi+Cellular), there are also icons for your cell signal and 3G/4G/LTE/EDGE/GPRS networks.

Performing Basic Operations

Now that you know where (almost) everything is on your iPad, it's time to learn some of the most common operations necessary to get things done.

View and Respond to Alerts

When your iPad or a specific app has something to tell you, you see an alert pop up onscreen. Depending on your settings and the importance of the alert, it may appear briefly at the top of the screen and then fade away, or it might remain in the center of the screen until you take some necessary action. Some alerts also appear on your iPad's Home or Search screens.

① Most alerts appear at the top of the current screen.

② Pull down on the alert to do something with it. Note that available options differ from app to app; for example, pulling down on a Messenger notification gives you options for Reply, Thumbs Up, Mute, and View Message. Pulling down a Mail notification gives you options for Trash, or Mark as Read.

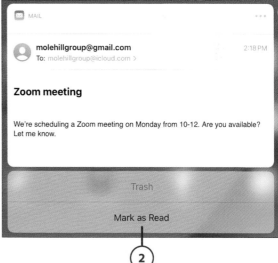

Configure Alerts

To determine which alerts you see (and how you see them), turn to Chapter 3.

Display the Control Center

When you want instant access to key system settings, such as screen brightness and volume, open your iPad's Control Center.

1. Swipe from the top-right corner of the screen toward the middle to display the Control Center.

2. Tap any card to turn on or off that control.

3. Tap and drag any slider to adjust that control.

4. Press and hold a card (this is called a *long press*) to display additional options for a given control.

5. For example, when you long press the Camera card, you see additional options for taking a selfie, recording a video, and so forth.

6. Tap outside the expanded panel to return to the normal Control Center. Close the Control Center by tapping on the background screen outside the controls. (You can also close the Control Center by pressing the Home button if your iPad has one.)

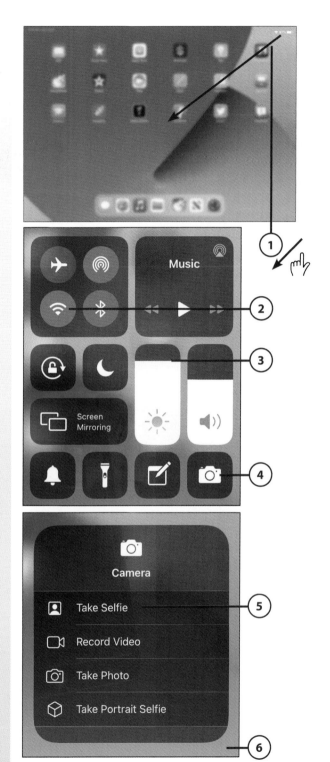

Adjust the Brightness

As noted, most of the more immediate iPad settings are accessed via the Control Center. You can control the device's screen brightness control via the Control Center.

① From within the Control Center, touch and drag the Brightness slider up to increase screen brightness.

② Touch and drag the Brightness slider down to decrease screen brightness.

Conserve Battery Usage
Lower the screen brightness to conserve the battery.

Activate Night Shift

Night Shift mode subtly adjusts screen colors so that they're warmer and easier to see at night. You can set a schedule for activating Night Shift, or simply go into Night Shift mode manually. (To configure Night Shift settings, tap Do Not Disturb on the Settings page.)

① From within the Control Center, long press the Brightness card to bring up the larger control.

(2) Tap the Night Shift icon to turn on Night Shift mode. (The first time you enable this feature, you are prompted to set the Night Shift schedule.) Tap the icon again to return to normal screen colors.

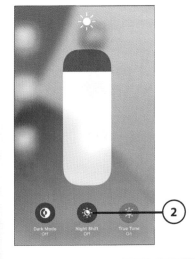

Change the Volume Level

Use the volume controls on the right side of your iPad to raise and lower the sounds you hear from the iPad's speakers.

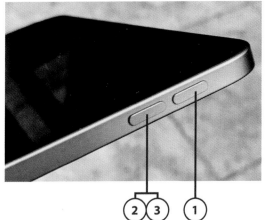

(1) Press the volume up button on the side of the iPad to increase the device's volume level.

(2) Press the volume down button on the side of the iPad to decrease the device's volume level.

(3) Press and hold the volume down button to temporarily mute the sound. Press the volume up button to increase the sound again.

(4) Alternatively, you can control the sound from the Control Center. Just open the Control Center and press and drag the Volume control up to increase the volume, or press and drag the control down to decrease the volume level.

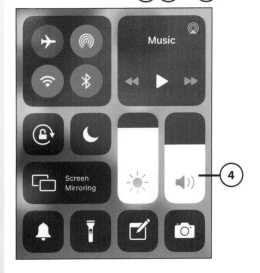

Rotate the iPad

As previously noted, you can use your iPad in either vertical (portrait) or horizontal (landscape) modes. The Home screen and most apps automatically rotate and adjust their displays to optimize how they work in either mode.

1. To switch from horizontal to vertical mode, simply rotate the iPad 90 degrees left or right.

2. The Home screen automatically adapts to the new orientation.

Activate Do Not Disturb Mode

There may be times when you don't want to be disturbed by alerts and notifications or by requests to video chat via FaceTime. When you don't want to be bothered by these and similar requests, activate your iPad's Do Not Disturb mode. When Do Not Disturb is activated, you won't see or hear notifications, alerts, and system sound effects. (Audio from music, movies, and TV shows is *not* muted.)

① From within the Control Center, tap the Do Not Disturb button to mute all system alerts and notifications.

② Tap the button again to return to normal operation.

Enter Text with the Onscreen Keyboard

Many applications let you (or even require you to!) enter text onscreen. You might be writing a note or memo or browsing a web page that asks you to enter information into an onscreen form.

When you need to enter text onscreen, use your iPad's onscreen keyboard. In most instances, the keyboard appears automatically, on the bottom half of the screen, when you tap within a form field or document.

1. Tap within a document or form field to display the onscreen keyboard.

2. Tap a key on the keyboard to enter that character onscreen.

3. Tap the Shift key to enter a capital letter. Double-tap the Shift key to engage caps lock.

4. Tap the Number key to display numbers and special characters.

5. Tap to enter a number or special character.

6. Tap the Emoji key to display and enter an emoji.

7. Tap a tab to display different types of emoji.

8. Tap to enter an emoji.

9. Tap the ABC key to return to the normal alphanumeric keyboard.

10. Tap the Keyboard key to hide the onscreen keyboard.

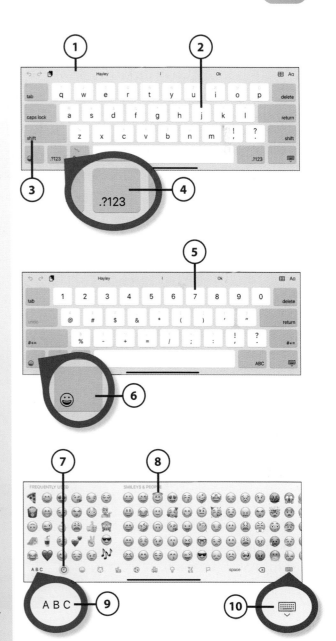

Floating Keyboard

To make the keyboard "float" over your document, pinch your fingers on the keyboard. This shrinks the keyboard, and you can move it around the screen.

>>>Go Further

PREDICTIVE KEYBOARD

Your iPad uses what is called a *predictive keyboard*, meaning that it tries to figure out what you're typing so it can enter the word for you. You see the recommended words at the top of the keyboard; tap a word to insert it into the form or document.

The predictive keyboard lets you type without having to completely enter long words, and it can help you type faster. It also helps you avoid common spelling mistakes.

If you don't like the suggestions made by the predictive keyboard, you can turn off the predictive part of it. From the Settings screen, select General; then select Keyboard. On the Keyboards screen, tap "off" the Predictive switch.

Copy and Paste Text

Whether you're using your iPad to write long letters or short Facebook posts, it's handy to be able to copy and paste text from one location to another.

(1) Double-tap a word to highlight that word.

(2) Tap and drag the starting and/or end points to select more or less text.

(3) Tap Cut to cut this selection (move it to another location).

(4) Tap Copy to copy this selection (duplicate it in another location).

(5) Move to where you want to paste the selected text and then double-tap the screen. This displays a different menu of options. (The available options differ from app to app, too.)

(6) Tap Paste.

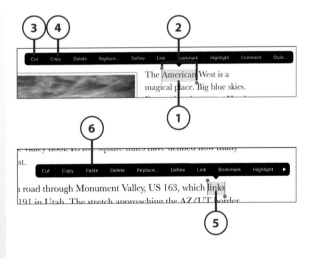

Set an Alarm

Your iPad includes a Clock app that lets you set a timer or an alarm. It also functions as a stopwatch.

If you want to use your iPad as a (very expensive!) alarm clock, use the Clock app to set an alarm.

1. Tap the Clock icon to open the Clock app.

2. Tap Alarm at the bottom of the screen.

3. Switch "on" any existing alarm to turn it on.

4. Tap + to create a new alarm.

5. Use the clock control to set the alarm time.

6. To have the alarm repeat on later days, tap Repeat and select which day(s) you want to use it.

7. Tap Label to create a name for this alarm.

8. Tap Sound to select the sound you want to hear when the alarm goes off.

9. Tap "off" the Snooze control if you don't want to be able to snooze through the alarm.

10. Tap Save. The alarm is created and activated.

Turn Off an Alarm

When an alarm sounds, turn it off by pressing the iPad's Home button. On an iPad without a Home button, tap the alarm notification to turn off the alarm.

Onscreen Switches

iPadOS uses a variety of onscreen switches to activate and deactivate various settings and functions. Just tap the switch to change its condition from "off" to "on," or vice versa. An "on" switch is to the right with a green background; an "off" switch is to the left with a light gray background.

Set a Timer

There are times when you need to set a timer—when you're cooking in the kitchen, for example, or putting the grandkids in a timeout. You can use the Clock app for all your timer needs.

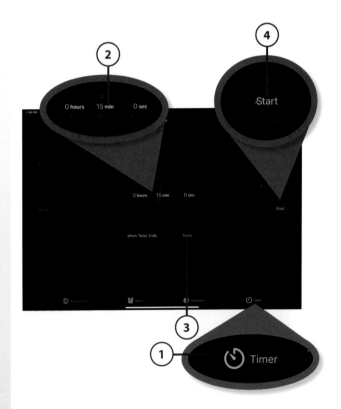

1. From within the Clock app, tap Timer.

2. Use the timer controls to set the length of the timer.

3. Tap When Timer Ends to set the sound you hear when the timer goes off.

4. Tap Start to start the timer.

Turn Off the Timer

When the Timer ends, press the Home button or the timer alert to turn it off.

Use the Stopwatch

The Clock app also lets you use it as a stopwatch, so that you can time any ongoing event.

1. From within the Clock app, tap Stopwatch.

2. Tap Start to start the stopwatch.

3. Tap Stop to stop the stopwatch.

4. The elapsed time is displayed onscreen.

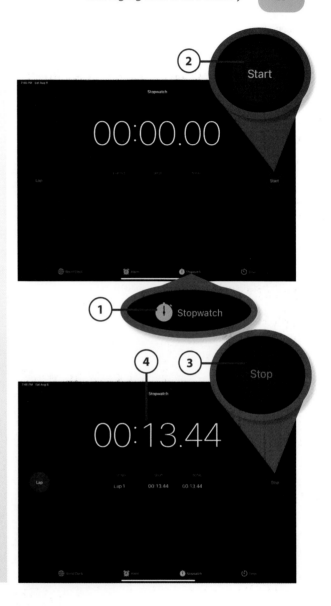

Managing Your iPad's Battery

Your iPad is powered by an internal battery that recharges when you connect it to the included power charger. How long the internal battery lasts before needing to be recharged depends on a lot of different factors—how you're using the device, which apps you're running, how bright you've set the screen, and so forth.

Recharge the Battery

Use the included power adapter to charge your iPad when you're not using it—although you can continue to use the device when charging.

(1) Connect the Lightning or USB-C connector on the supplied cable to the port on the bottom of the iPad.

It Works Either Way

One nice thing about both Lightning and USB-C connectors is that they're symmetrical; you can plug the cable in right side up or upside down; it's all the same. No more twisting and turning to fit cables into connectors!

(2) Connect the USB connector on the cable to the USB port on the power adapter.

(3) Connect the power adapter to any power outlet. Your iPad should fully charge in about 5 to 7 hours—longer if you're using it while charging.

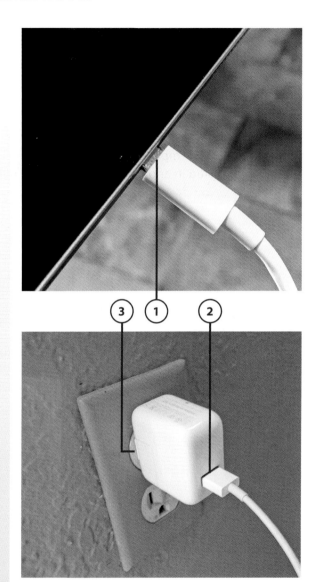

Monitor Battery Usage

The battery icon on the right side of the status bar indicates how much charge you have left. When the charge gets too low, the status bar displays the image of a nearly empty battery, and you need to plug in and charge up your iPad before you can continue to use it.

If you want to know which apps are using the most battery power, you can do that.

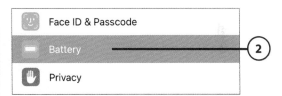

1 Tap the Settings icon to open the Settings screen.

2 Tap Battery in the left column.

3 The right side of the screen displays graphs for Battery Level and Activity. Examine these graphs to learn more about your iPad's battery use.

4 Tap to display usage for the last 24 hours or the last 10 days.

5 If you scroll down, you see a list of those apps and system functions that have used the most battery power, in descending order. Tap any item to view how many minutes this and other items have been used.

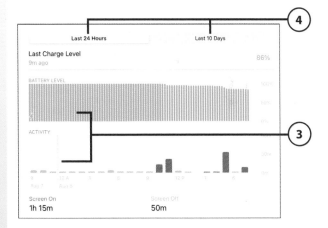

In this chapter, you learn how to personalize various aspects of your iPad experience.

→ Personalizing the Lock and Home Screens
→ Configuring Device Settings
→ Managing System Settings
→ Managing Your Screen Time

3

Personalizing the Way Your iPad Looks and Works

Every iPad looks and acts the same right out of the box. But that doesn't mean you can't configure your iPad to your tastes.

Want a different background picture? You can do that. Want a brighter or dimmer screen? You can do that, too. Don't like the way you receive notifications onscreen, and from which apps? Then change the notification settings.

Truth is, there's a lot you can do to personalize your iPad. This chapter walks you through managing your screens, backgrounds, and more detailed settings. Change is coming!

Personalizing the Lock and Home Screens

One of the most visible things that people like to change is the look of the iPad screens. You can change the background wallpaper of the Lock and Home screens as well as the way app icons are arranged onscreen.

Change the Wallpaper

The default wallpaper you see on your iPad's Home and Lock screens is pretty enough, but you can probably find something more to your liking. You can choose different wallpaper for the Lock and Home screens or use the same wallpaper for both.

① Tap the Settings icon to display the Settings screen.

② In the left column, scroll down and tap Wallpaper.

③ Your current wallpaper is displayed on thumbnails of your Lock (left) and Home (right) screens. Tap Choose a New Wallpaper.

④ Tap Dynamic to select one of Apple's supplied dynamic (moving) wallpapers. *Or…*

⑤ Tap Stills to select one of Apple's supplied still photos.

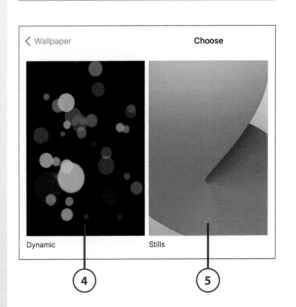

6 Tap to select one of the available choices. (Light mode version is on the left; dark mode version is on the right.) You see the wallpaper applied.

Dark Mode

Dark mode displays a darker background on most screens and also affects how backgrounds display. Learn more about dark mode in the "Choose Light or Dark Mode" section, later in this chapter.

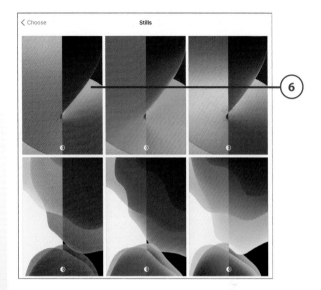

7 Tap Set to keep this wallpaper or Cancel (not shown) to keep looking.

8 If you tapped Set, tap Set Lock Screen to set this picture for only your Lock screen. *Or…*

9 Tap Set Home Screen to set this picture for only your Home screen. *Or…*

10 Tap Set Both to use this picture as the background for both your Lock and Home screens.

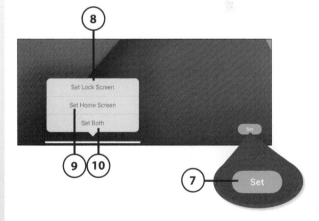

Dim in Dark Mode

If you want your wallpaper to dim when your iPad is in dark mode, go to the Wallpaper settings screen and tap "on" the Dark Appearance Dims Wallpaper switch.

Change the Size of App Icons

You can choose to display either small or large app icons on the Home screens. To display small icons (up to 30 icons per screen), you select the More icons option. To display large icons (up to 20 icons per screen), you select the Bigger option.

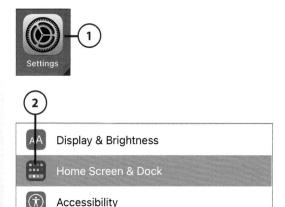

Apps

Learn more about the apps behind the app icons in Chapter 8, "Installing and Using Apps."

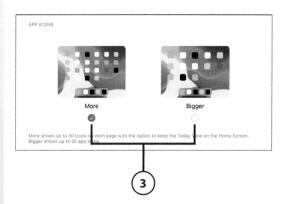

1. Tap the Settings icon to display the Settings screen.

2. Tap Home Screen & Dock on the left side.

3. In the App Icons section, select either More (for smaller icons) or Bigger (for larger icons).

Arrange App Icons

Which icons appear where on your Home screens is totally up to you. You can easily move icons from one position to another—even from one screen to another.

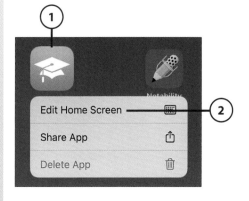

1. Press and hold the icon you want to move.

2. Select Edit Home Screen from the pop-up menu. All the icons onscreen start to jiggle.

3 You also see a little X next to each icon, which means you're in editing mode. Tap the X to delete an app.

4 To move an icon, drag it to a new position. The other icons rearrange themselves to make room. Lift your finger when the icon is in place.

5 Press the Home button or, if your iPad doesn't have a Home button, tap Done to return to normal screen mode.

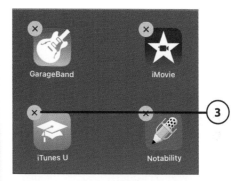

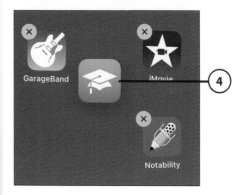

Screen to Screen

To move an icon to a different Home screen, drag the icon to the side of the screen until the adjacent screen appears. You can then "drop" the icon to a new position on this screen. If you try to drag an icon to a screen that's already full, it pushes the last icon (on the bottom right) to the next screen.

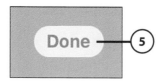

Add Icons to and Remove Icons from the Dock

The Dock is that area at the bottom of every screen that holds icons for your favorite and recently used apps. Your favorite apps are docked on the left side; recently used apps display on the right side of the Dock.

You can remove any of the current app icons or add icons for other apps. In fact, you can populate the Dock with more than a dozen icons.

If you stock up the Dock with icons for the apps you use most, you won't have to move back to the Home screen as often. Because the Dock hovers on top of most open apps (and you can redisplay it simply by swiping up from the bottom of the screen), you can do all your app launching from the Dock rather than from the Home screen.

1 Add an icon to the Dock by dragging it from any screen down to
the left side of the Dock.

2 Rearrange icons on the Dock
by dragging any icon to a new
position.

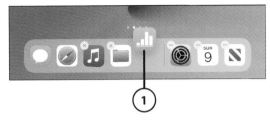

3 Remove an icon from the Dock by
dragging it from the Dock to any
Home screen.

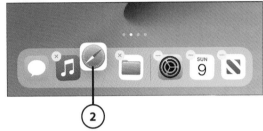

Organize Icons with Folders

If you have too many app icons
crowding too many screens, you can
organize those apps into folders. Each
folder can contain multiple apps; to
open a folder, tap it and then tap the
appropriate app inside.

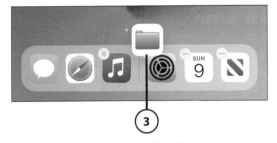

1 Create a new folder by dragging
one icon on top of another. This
creates the folder, gives it a name,
and displays it full screen.

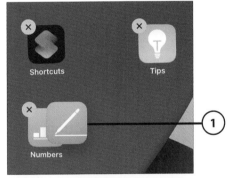

2 Rename the folder by pressing
the folder and selecting Rename.
You can then use the onscreen
keyboard to edit or enter a new
name. (You can also rename an
open folder by tapping the folder
name.)

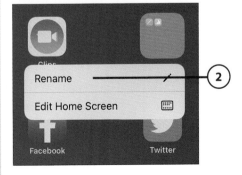

(3) Remove an app from a folder by opening the folder and then dragging the icon outside of the folder.

(4) Tap anywhere outside the folder to close the folder and return to the desktop. The folder is displayed in position there. (You can drag the folder to move it to another position if you like.)

(5) Add another app to this folder by dragging the app icon on top of the folder icon.

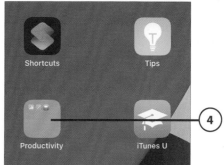

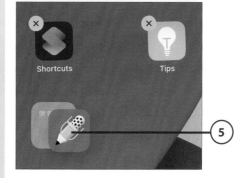

>>>Go Further

ORGANIZE BY TYPE

I find it useful to organize my lesser-used apps into folders. I want to have immediate access on the Home screen to those apps I use all the time; I put those apps I use less regularly into folders so they're not cluttering my screens.

When I use folders, I try to put similar apps together. For example, if I have a half dozen or so games, I might create a folder labeled Games to hold them all. Similarly, I might group my music, video, and photos apps into a folder labeled Lifestyle, or the Numbers, Pages, and Keynote apps into a Productivity folder.

The point is to group similar apps in folders where I can quickly and easily find them. I've found that creating a folder labeled Misc is the best way to "lose" apps on my device!

Display Today View

Today View is a collection of "widgets" that pulls in from the left side of any Home screen. A widget is a small panel that displays specific information, such as news, weather, appointments, and the like, often from related apps.

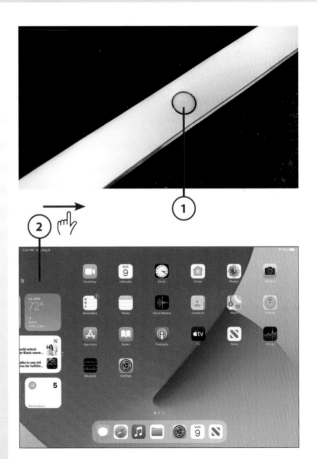

Today View from the Lock Screen

You can also display Today View before you unlock your iPad. Just swipe to the right from the Lock screen, and you'll see all your Today View notifications and widgets.

1. Press the Home button or swipe up from the bottom of any screen to display the Home screen.

2. Swipe in from the left side of the screen to display Today View.

(3) Tap any widget or notification to view more details or to open the corresponding app.

(4) To hide the Today View panel, swipe it to the left.

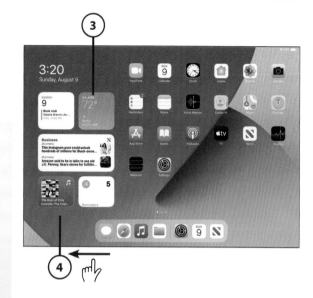

Put Today View on the Home Screen

You can pin the Today View panel to the Home screen when it's in horizontal (landscape) mode. (In vertical [portrait] mode, there's simply not enough width to display these widgets.) Today View appears on the left side of the screen, and your normal app icons appear on the right.

Here's how to pin Today View to the Home screen. (Remember, you must hold your iPad horizontally for this to work.)

(1) Swipe right on the Home screen to display Today View.

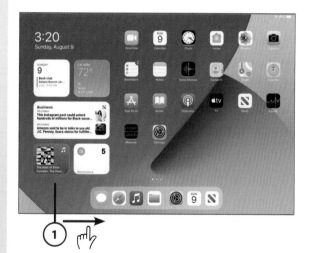

(2) Swipe up through Today View Widgets and then tap the Edit button at the bottom.

(3) Tap "on" the Keep on Home Screen switch.

(4) Tap Done when done.

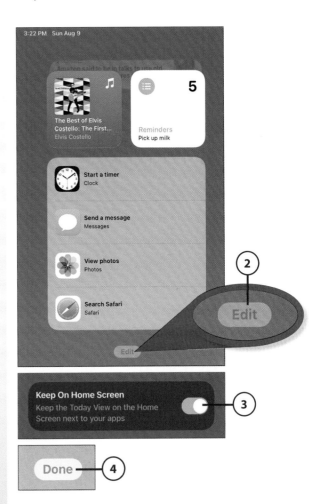

Customize Widgets in Today View

You can personalize which widgets display in Today View. This makes Today View more useful for you.

(1) Scroll to the bottom of Today View and tap Edit.

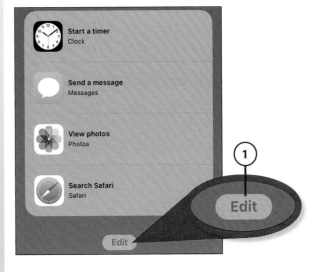

2 To move a widget to another position, press and drag the widget to its desired position.

3 Tap the − (minus) icon to remove a widget from the Today screen.

4 To add a new widget for an app installed on your device, scroll to the bottom of the Today View and tap Customize.

5 To add one of the widgets, tap the + (plus) icon.

6 Tap Done.

7 Even more widgets are available in the Widget Gallery. Tap the + at the top of Today View to display the Gallery.

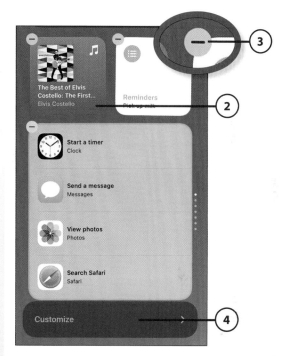

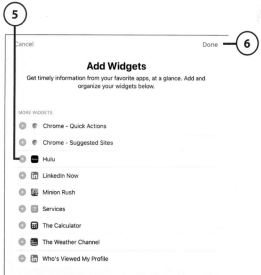

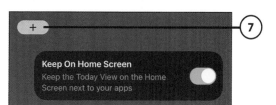

8   Tap the widget you wish to add.

9   Swipe right and left through the different sizes and options for this widget. Stop on the page you want to add.

10   Tap Add Widget to add the widget.

11   Tap Done when done.

Smart Stack

In iPadOS 14, Apple introduces the Smart Stack widget. The Smart Stack is a collection of widgets that use on-device intelligence to display the right widgets at the right time, based on time of day, location, and your activity. For example, the Smart Stack might display News in the morning, Calendar events throughout the day, and your Activity summary in the evening. You add the Smart Stack widget from the Widget Gallery.

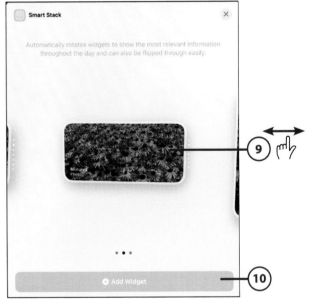

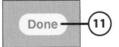

Customize the Control Center

When you swipe from the top-right corner of any screen down toward the middle of the screen, you display the Control Center. A number of useful controls are displayed by default in the Control Center; you can personalize the Control Center by adding additional controls.

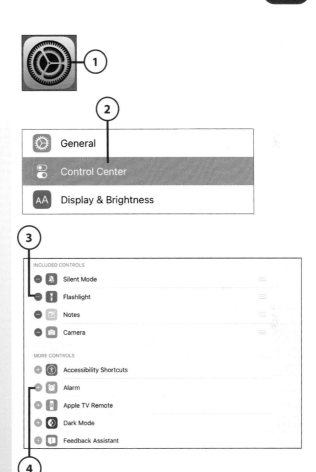

1 Tap the Settings icon to display the Settings screen.

2 In the left column, tap Control Center.

3 To remove a control from the Control Center, tap the red – next to that control's name and then tap Remove.

4 To add a new control to the Control Center, tap the green + next to that control's name.

Configuring Device Settings

There are many other settings you can configure to better personalize your iPad. You configure all these settings from the iPad's Settings page.

Display the Settings Page

The Settings page is organized into a series of tabs, which are accessible from the left column. Tap a tab to view the individual settings of a given type.

(1) Tap the Settings icon to open the Settings screen.

(2) Tap a tab on the left to display related settings.

Configure Notifications

Many apps display notifications for various types of events. The Mail app, for example, can display notifications when you receive new emails; the News app can display notifications with recent news headlines.

You can determine what types of notifications you receive on an app-by-app basis. For that matter, you can opt to turn off all notifications for specific apps, thus reducing the number of annoying notifications you have to deal with.

Different Notifications
Different apps have different notification options. This how-to discusses some of the more common notification options.

---

(1) From the Settings screen, tap Notifications in the left column to display the Notifications tab.

(2) To change when previews for notifications are displayed, tap Show Previews.

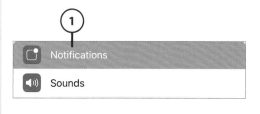

(3) Tap Always to always display notification previews. Tap When Unlocked to display notification previews only when your iPad is unlocked. Tap Never to never display notification previews.

(4) Tap Notifications to return to the previous screen.

(5) To configure notifications for a specific app, tap that app.

(6) Tap "on" the Allow Notifications switch to display notifications from this app. Tap "off" this control to disable all notifications from this app.

(7) Show notifications from this app on the Lock Screen by tapping to select Lock Screen. Deselect this option if you don't want notifications to show on the Lock Screen.

(8) Show notifications from this app in the Notification Center by tapping to select Notification Center. Deselect this option if you don't want notifications to show in the Notification Center.

(9) Display alerts as banners by tapping to select Banners and then selecting a banner style. Temporary banners go away automatically after a few seconds. Persistent banners stay onscreen until you interact with them.

(10) Tap Sounds to choose a notification sound for this app or turn off notification sounds. (Not all apps have notification sounds.)

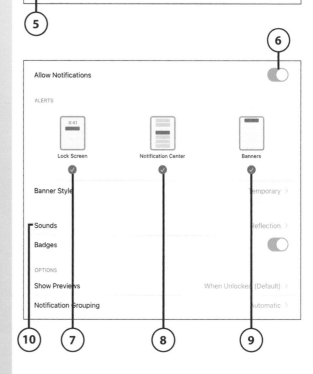

11 Some apps display a number on top of the icon to indicate the number of notifications or actions pending. These numbers are called *badges*. (For example, the icon for the Mail app displays a red "2" if there are two unread emails in the inbox.) Turn on this numbering by tapping "on" the Badges switch.

12 To show previews from this app (not available with all apps) in the notifications, tap Show Previews and then select what type of preview: Always, When Unlocked, or Never.

13 To change how similar notifications are grouped, tap Notification Grouping and then select Automatic (your iPad groups them in a logical order, by topic), By App (all notifications from a given app are grouped together), or Off.

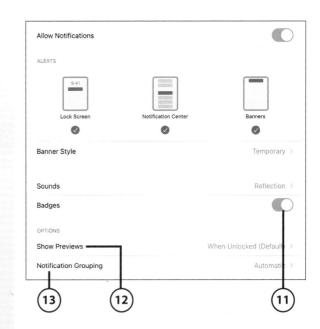

Choose Light or Dark Mode

There are two different modes for your iPad display—Light mode and Dark mode. Light mode presents most screens with the traditional light background and black text. Dark mode displays most screens with a black background and white text, which some people find easier to view at night.

1 From the Settings screen, tap Display & Brightness in the left column.

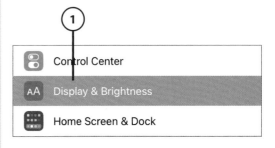

2 In the Appearance section, tap Light to select Light mode.

3 Tap Dark to select Dark mode.

4 Tap "on" the Automatic switch to have your iPad switch to Dark mode at sunset and Light mode at sunrise. (You can then tap Options to choose a custom schedule for Light and Dark modes.)

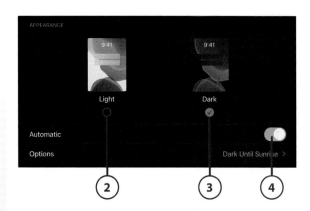

Configure the Display and Brightness

You can adjust how bright your iPad's display appears, along with several other display-related settings.

1 From the Settings screen, tap Display & Brightness in the left column.

2 Drag the Brightness slider to the left to make the screen less bright or to the right to make it brighter. (Remember, the brighter the screen, the faster it drains your battery.)

3 To have your iPad adapt its display for different ambient lighting conditions, tap "on" the True Tone switch. (True Tone is not available on all models.)

4 To let your iPad automatically shift the display colors to be warmer (more reddish) after dark, tap Night Shift. You see the Night Shift panel.

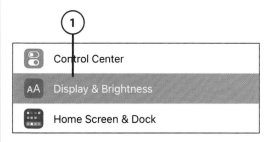

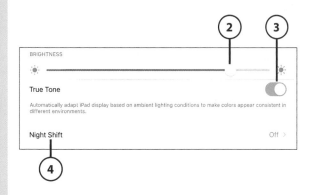

5 Schedule when Night Shift is enabled by tapping "on" the Scheduled switch and making a selection.

6 Manually enable Night Shift from now to daylight tomorrow by tapping "on" the Manually Enable Until Tomorrow switch.

7 Adjust how warm the Night Shift colors appear by dragging the Color Temperature slider to the left (less warm) or right (warmer).

8 Tap Display & Brightness to return to the Display & Brightness screen.

9 Change the amount of idle time that must elapse before your iPad automatically locks by tapping Auto-Lock and making a new selection (2 minutes, 5 minutes, 10 minutes, 15 minutes, or never).

10 If you're using your iPad with a smart cover, tap "on" the Lock/Unlock switch to let the cover automatically lock and unlock your iPad.

11 To make the text for icons and other system elements bold (and thus, perhaps, easier to read), tap "on" the Bold Text switch.

12 Tap Text Size to change the size of the text you see onscreen.

13 Drag the slider to the left to make the text smaller or to the right to make the text larger.

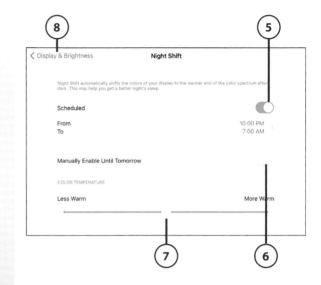

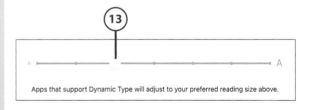

Change System Sounds

You can select different sounds for different system actions, such as sending an email or making a Facebook post. You also can opt to hear sounds when your iPad locks or when you tap the keys of the onscreen keyboard.

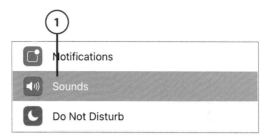

(1) From the Settings screen, tap Sounds in the left column.

(2) To keep sudden loud sounds from damaging your hearing when you're listening via earbuds or headphones, tap Reduce Loud Sounds; then, on the next page, tap "on" the Reduce Loud Sounds switch.

(3) Drag the Ringer and Alerts slider to the left to decrease the volume; drag to the right to make them louder.

(4) To control these system sounds with your iPad's physical volume buttons, tap "on" the Change with Buttons switch. (Otherwise, the volume buttons control only the volume of music and videos you're listening to or watching.)

(5) Change the sound for any specific system function by going to the Sounds section, tapping that function, and selecting a new sound.

(6) Tap "off" the Keyboard Clicks switch to turn off sounds when you tap your iPad's onscreen keyboard. (Some people find all these clickety-clacks annoying when they type.)

(7) Tap "off" the Lock Sound switch to turn off sounds when you lock or unlock your iPad.

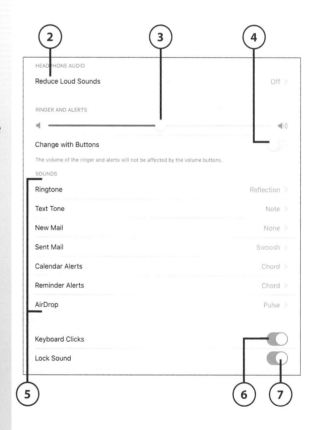

Configure Individual Apps

Most of the apps installed on your iPad have individual settings you can configure. Use the Settings screen to adjust settings for any app.

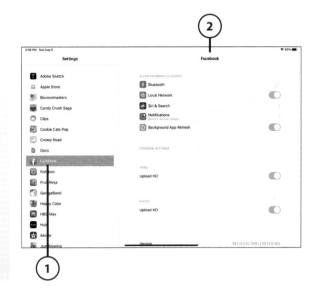

(1) From the Settings screen, scroll down to the list of apps in the left column and tap an app.

(2) That app's settings are displayed on the right side of the screen.

Managing System Settings

I've covered some of the most used settings you may want to configure on your iPad. But there are more settings available on the General tab that let you fine-tune various system options—and show you information about your device.

Display and Manage General Settings

The General tab on the Settings page is home to a variety of settings that affect all the apps on your iPad—as well as the iPad itself.

(1) From the Settings screen, tap General in the left column to select the General tab.

2 Tap About to see information about your device—how many items of various types you have stored, total capacity, available storage space, model and serial number, and more.

3 Tap Software Update to manually force any available software update.

Automatic Updates

Software updates automatically install when you have the Automatic Updates switch on; manual updates are not typically needed.

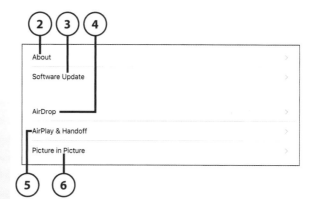

4 Tap AirDrop to configure your iPad's AirDrop feature, which lets you instantly share contacts and other data with nearby Apple devices.

5 Tap AirPlay & Handoff to activate the ability to start work on one device and resume that work on another device (Handoff), as well as to automatically cast media from your iPad to your TV (AirPlay).

6 Tap Picture in Picture to automatically continue videos and FaceTime calls in a small PIP window when you switch to the Home screen or other apps (activated by default).

(7) Tap iPad Storage to view how much storage space you've used.

(8) Some apps automatically refresh themselves in the background. To turn this off, tap Background App Refresh.

(9) Tap Date & Time to set your iPad's date and time.

(10) Tap Keyboard to configure various options for the onscreen keyboard.

(11) Tap Fonts to select any custom-installed fonts for your iPad.

(12) Tap Language & Region to reset the language and region for your iPad.

(13) Tap Dictionary to select which dictionary your iPad uses when checking text you enter.

(14) Tap VPN to configure your iPad to use a virtual public network.

(15) Tap Profile to view your iPad's configuration profile.

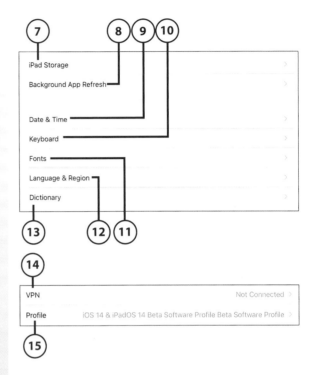

Managing Your Screen Time

The Screen Time feature enables you to view and manage the amount of time you spend using your iPad. Screen Time can help you (or anyone else using your iPad) to spend less time with your eyes on the screen and more time interacting with real people in the real world.

View Your Screen Time

Perhaps the most useful feature of Screen Time is the ability to see how much time you've spent using your iPad—and on what activities you're spending that time.

(1) From the Settings screen, tap Screen Time in the left column.

(2) The Daily Average panel at the top of the Screen Time tab displays the amount of time you've used your iPad over the past several days.

(3) Tap See All Activity to view more information.

(4) Tap Week to view your activity for the week, or tap Day to view your activity today.

(5) The Most Used panel displays information about your most used apps and websites. Tap an app to view more information about that item.

(6) By default, the Most Used panel displays information about apps and websites. Tap Show Categories to show this information grouped by type of activity.

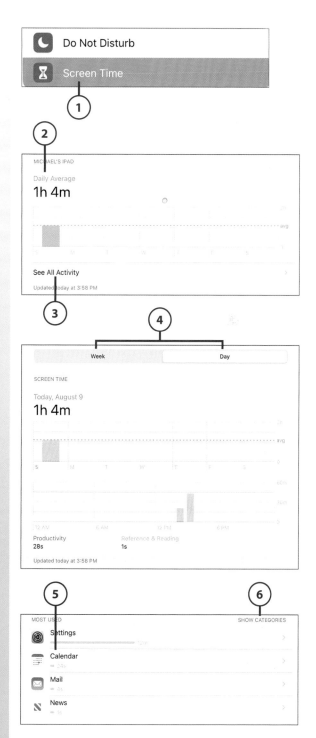

(7) The Pickups panel displays how many times you've picked up your iPad and when. It also shows during what time spans you've picked up your iPad the most.

(8) The Notifications panel displays how many notifications you've received, when, and of what type.

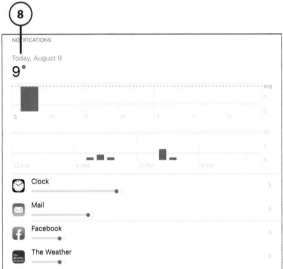

Configure Downtime

If you think you're using your iPad too much, use Screen Time to schedule regular downtime when you won't be allowed to use your device—except for those apps you select to be available. For example, you may want to still send and receive messages during your downtime.

(1) From the Screen Time tab, tap Downtime.

(2) Tap "on" the Downtime switch to enable the downtime feature. (Tap "off" this switch if you don't want any iPad downtime.)

(3) Tap Every Day (enabled by default) to have Downtime apply every day of the week.

(4) Tap Customize Days to select specific days of the week for Downtime to apply.

(5) Tap From and enter the time you want your downtime to start.

(6) Tap To and enter the time you want your device to be available again.

(7) Tap Screen Time to return to the Screen Time tab.

(8) Tap Always Allowed.

(9) Apps allowed during downtime are displayed in the Allowed Apps panel. Tap the red − to remove an app from the allowed list.

(10) To add another app to the allowed list, tap the green + next to it in the Choose Apps panel.

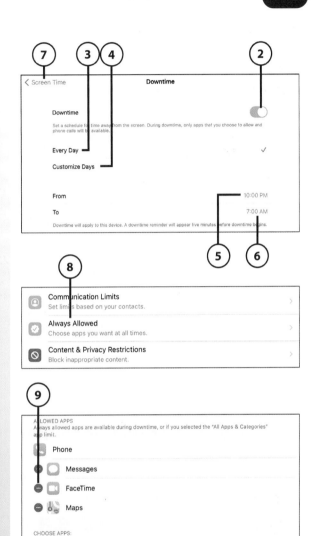

Set App Limits

The Screen Time app also enables you to set time limits for specific apps. Once you set a limit, you will be able to use that type of app only for that specific amount of time each day.

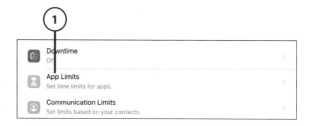

(1) From the Screen Time tab, tap App Limits.

(2) Tap "on" the App Limits switch to activate this feature. (This switch only appears after you've added one or more limits.)

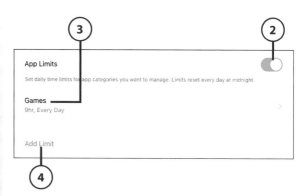

(3) Any app limits you've previously set are displayed here. Tap a limit to edit or delete that limit.

(4) Tap Add Limit to add a new app limit.

(5) Tap the category you want to limit, or tap to expand a category and then tap specific apps.

(6) Tap Next.

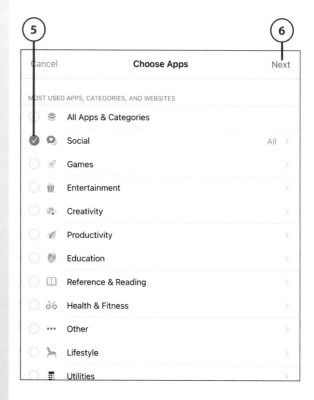

(7) Set the amount of time you want to use that type of app during a 24-hour period.

(8) To set custom app limits for different days of the week, tap Customize Days.

(9) Tap Add to add this limit.

Edit or Delete App Limits

To change or delete an app limit, select it on the App Limits screen. Tap "off" the App Limit switch to turn off the limit. Tap Time to change the length of the time limit. Tap Delete Limit to completely get rid of this app limit.

Block Specific Content

You can also use Screen Time to block any content you want—including inappropriate content.

(1) From the Screen Time tab, tap Content & Privacy Restrictions.

(2) Tap "on" the Content & Privacy Restrictions switch.

(3) By default, everything is allowed. To block specific content, tap that item.

(4) Tap Don't Allow Changes (or Don't Allow or just "off," depending on the available options) for the specific item you want to block.

Blocking App Store Purchases

If you don't want your iPad used to purchase new apps (or make in-app purchases), tap iTunes & App Store Purchases, tap Installing Apps or In-App Purchases, and then tap Don't Allow. If you want to always require a password before a purchase, tap iTunes & App Store Purchases, go to the Require Password panel, and then tap Always Require.

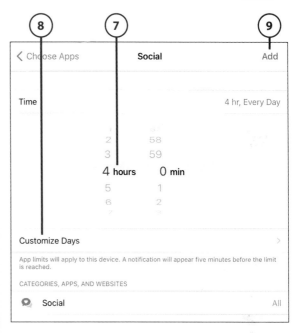

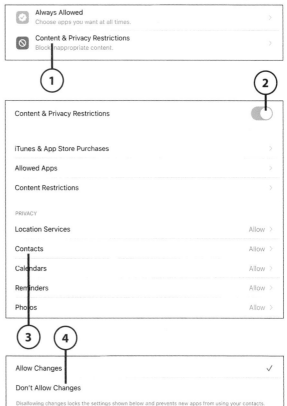

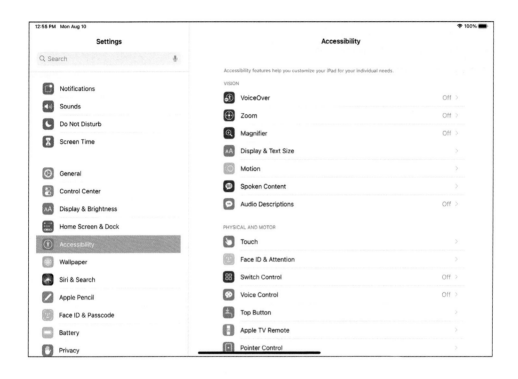

In this chapter, you explore how to configure your iPad's accessibility options—and make it easier to use.

→ Making the iPad Easier to See
→ Making the iPad Easier to Hear
→ Making the iPad Easier to Operate

4

Making Your iPad More Accessible

Because even the largest iPad has a relatively small screen, some of us may have trouble seeing what's onscreen or tapping where we need to tap. Others may have difficulty listening to sounds coming from the iPad's equally small speakers.

Fortunately, several settings on your iPad can make it easier to use. I discuss them in this chapter.

Making the iPad Easier to See

Let's start with those features that can make things displayed onscreen easier to see. Many of us have some degree of vision loss, even if it's just the need for reading glasses to see fine print. Given the relatively small size of text on the iPad screen (especially on many web pages), you may want to avail yourself of these features that make onscreen text easier to read.

Read the Screen with VoiceOver

Probably the most useful accessibility feature for those with vision difficulties is Apple's VoiceOver. VoiceOver describes out loud any screen element or text. Just touch an item or text selection and VoiceOver either reads it or tells you about it. In some instances, VoiceOver even tells you how to use a given item— "double-tap to open," for example.

When VoiceOver is activated, all you have to do is touch the screen or drag your finger over an area to hear information about what's onscreen. When you go to a new screen, VoiceOver plays a sound and then selects and reads the first item on the new screen.

Note that because VoiceOver requires you to tap the screen to "read" an item, most traditional touch gestures change when you've activated VoiceOver. Where you would normally use a single-tap, for example, you now use a double-tap.

Other Screen-Reading Options

Although VoiceOver is the best and most full-featured screen-reading option on your iPad, other options are available. Go to the Accessibility screen and tap Spoken Content to enable the Speak Selection (for reading text), Speak Screen (for reading screen elements), and Typing Feedback (for reading words while you type) options. Give them a try; they're less functional but also less intrusive than VoiceOver.

---

(1) To activate VoiceOver, go to the Settings screen and select the Accessibility tab.

(2) Tap VoiceOver.

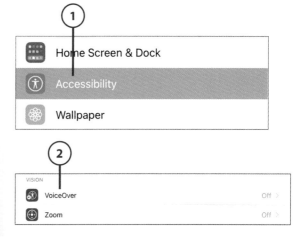

3. Tap "on" the VoiceOver switch. (Alternatively, press and hold the Home button and tell Siri, "Turn VoiceOver on.")

4. To "read" any onscreen item, touch it. (The item is now surrounded by a black border—the VoiceOver cursor.)

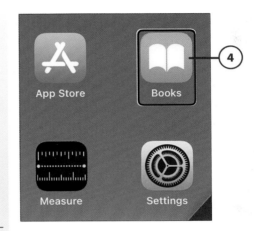

5. To activate any operation normally done by single-tapping, instead select the item and then double-tap the screen. (Not shown.)

6. To activate any operation normally done by double-tapping, instead triple-tap that item. (Not shown.)

Turn It Off

To deactivate VoiceOver, return to the VoiceOver screen and double-tap the VoiceOver switch.

>>>Go Further

VOICEOVER GESTURES

When you're using VoiceOver, how you use your iPad changes in subtle ways. Because basic tapping or touching now "reads" that onscreen item, you need to use different touch gestures to do just about anything onscreen.

The following table shows you some of the more common operations and their new VoiceOver-enabled touch gestures.

Operation	VoiceOver Gesture
Select and read an item	Single-tap
Select the next or previous item	Swipe right or left
Go to the Home screen	Slide one finger up from the bottom until you hear the second sound and then lift your finger

Operation	VoiceOver Gesture
Open the App Switcher	Slide one finger up from the bottom until you hear three sounds and then lift your finger
Open the Control Center	Slide one finger down from the top edge until you hear the second sound and then lift your finger
Open the Notification Center	Slide one finger down from the top until you hear the third sound and then lift your finger
Read all elements from the top of the screen	Two-finger swipe up
Read all elements from the current position	Two-finger swipe down
Stop or resume speaking	Two-finger tap
Dismiss an alert or go back to a previous screen	Move two fingers back and forth quickly in a Z pattern
Scroll one page at a time	Three-finger swipe up or down
Go to the next or previous page (including on the Home page)	Three-finger swipe right or left
Speak additional information about the selected item	Three-finger tap
Select the first item on the page	Four-finger tap at the top of the screen
Select the last item on the page	Four-finger tap at the bottom of the screen
Activate the selected item	Double-tap
Double-tap an item	Triple-tap
Initiate an action or pause an in-progress action	Two-finger double-tap
Change an item's label	Two-finger double-tap and then hold
Open the Item Chooser	Two-finger triple-tap
Mute or unmute VoiceOver	Three-finger double-tap

There are also a number of settings you can configure to personalize the way VoiceOver works for you. Just go to the Settings page, select the Accessibility tab, and then tap VoiceOver.

Braille Displays

Your iPad is fully compatible with most third-party braille displays. Use the braille display to control your iPad when VoiceOver is activated. Connect a braille display to your iPad via Bluetooth, and then go the Settings page, select the General tab, tap Accessibility, tap VoiceOver, and then tap Braille.

Magnify the Screen with Zoom

Having VoiceOver take over your iPad's screen may be too much if you have only mild vision difficulties. In this instance, you might need only to zoom in to an area of the screen to make it larger and more legible.

Many apps let you zoom in by double-tapping the screen or moving your fingers apart. (You zoom out by pinching your fingers together.) Apple also offers a dedicated Zoom feature that lets you magnify any screen.

(1) To activate Zoom, go to the Settings screen, select the Accessibility tab, and then tap Zoom.

(2) Tap "on" the Zoom switch to display the zoom box onscreen.

(3) Drag the bottom of the zoom box to move it around the screen. (Not shown.)

(4) Hide the zoom box by double-tapping the screen with three fingers. Double-tap with three fingers again to redisplay the zoom box. (Not shown.)

Invert Screen Colors

If you're having trouble distinguishing elements onscreen, it may be an issue of contrast. I've found that increasing the contrast between light and dark elements can help me read even very small type onscreen.

With this in mind, Apple enables you to invert your colors onscreen. With inverted colors, black text on a white background becomes white text on a black background, and other color highlights appear against that same black background.

(1) Go to the Settings screen, select the Accessibility tab, and then tap Display & Text Size.

(2) Tap "on" the Smart Invert switch to reverse-display colors except for images, video, and apps that use a dark color scheme. (This is the preferred option.)

(3) Tap "on" the Classic Invert switch to reverse all colors in all apps. (Return to your iPad's normal color scheme by returning to the Display & Text Size screen and tapping "off" whichever option you've previously enabled.)

Configure the Screen for Colorblind Users

If you're fully or partially colorblind, you simply can't see some color combinations. Fortunately, you can configure your iPad with color filters to make those otherwise hard-to-see colors pop.

1. Go to the Settings screen, select the Accessibility tab, and then tap Display & Text Size.

2. Tap Color Filters.

3. Tap "on" the Color Filters switch

4. Select the filter that works best for you.

5. Use the Color Intensity slider to adjust the intensity of the colors.

Make Onscreen Text Larger and Bolder

Another way to make an iPad screen more readable is to increase the size of the onscreen text or perhaps even bold that text.

1. Go to the Settings screen, select the Accessibility tab, and then tap Display & Text Size.

2. Make onscreen text bold by tapping "on" the Bold Text switch.

3. Increase the size of onscreen text by tapping Larger Text.

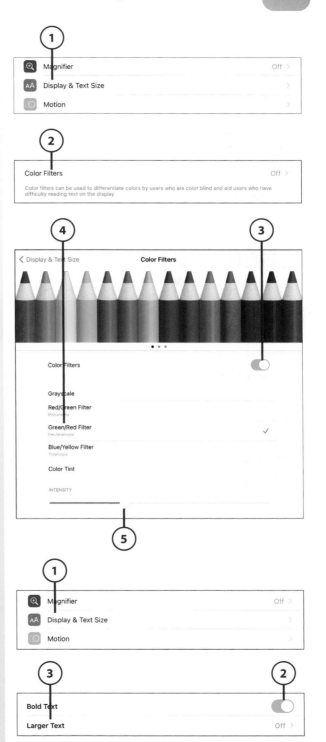

4 Tap "on" the Larger Accessibility Sizes switch.

5 Drag the slider to the right to make text larger. Drag the slider back to the left to reduce the size of the text.

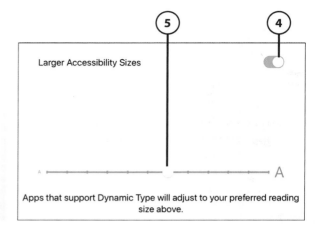

Making the iPad Easier to Hear

Whether you're a little hard of hearing or completely deaf, you'll be pleased to know that your iPad can accommodate various types of hearing devices, as well as compensate for hearing damage in one ear or the other.

Use a Hearing Aid with Your iPad

Many companies make "Made for iPad" hearing aids that are officially compatible with most iPad models. When you pair these hearing aids via Bluetooth, you can use the iPad to adjust their settings, stream audio directly to the hearing aids, and more. All you have to do is configure your iPad for the type of hearing aid you have. Start by opening the battery doors on your hearing aids, and then follow these steps.

1 Make sure Bluetooth is enabled on your iPad (tap Bluetooth from the Settings screen); then go to the Settings screen, select the Accessibility tab, and then scroll to the Hearing section and tap Hearing Devices.

2 Close the battery doors on your hearing aids and wait for them to appear in the Devices list. When the hearing aid appears, tap its name and respond to the pairing request. You can then adjust settings specific to your hearing aids from the Hearing Devices page.

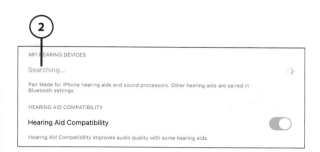

Use Mono Audio

If your hearing is weaker in one ear than the other, you might have trouble hearing sounds coming from either your left or right hearing aid. To compensate, you can mix all stereo sound to mono and have your iPad play back only mono sounds. From the Accessibility screen, go to the Hearing section, tap Audio & Visual, and then tap "on" the Mono Audio switch. You can then drag the slider to the left or right until you can best hear all sounds coming from your iPad.

Turn On Closed Captioning When Watching Videos

Many streaming video services and video player apps, such as Apple's Videos app, offer closed captioning for people with hearing impairments. You can typically enable closed captioning within the individual app or universally via your iPad's general settings.

1 Go to the Settings screen, select the Accessibility tab, and then scroll to the Hearing section and tap Subtitles & Captioning.

(2) Tap "on" the Closed Captions +
SDH switch.

(3) Tap Style to select a closed-cap-
tioned style.

Making the iPad Easier to Operate

If you have trouble performing the necessary touch gestures to operate your
iPad, there are some options available to you. You can fine-tune how various
gestures work, enable an assistive technology called AssistiveTouch, or just use
the Siri personal assistant to operate your iPad via voice commands.

Adjust the Touchscreen Display

Sometimes I find my iPad's touchscreen display to be a little too touchy.
If you find yourself tapping or swiping
things without meaning to, you can
adjust the sensitivity of the display.

(1) Go to the Settings screen, select
the Accessibility tab, and then
scroll to the Physical and Motor
section and tap Touch.

(2) Tap Touch Accommodations.

3 Tap "on" the Touch Accommodations switch.

4 Adjust how long you must touch the screen before that touch is recognized by tapping on the Hold Duration switch and then setting a new time. (The default is 0.1 second.)

5 If your single touches sometimes register as double touches, tap the Ignore Repeat switch and increase the time limit.

6 If you have trouble precisely tapping onscreen elements, enable Tap Assistance by tapping either Use Initial Touch Location or Use Final Touch Location. Tap Off to disable this feature.

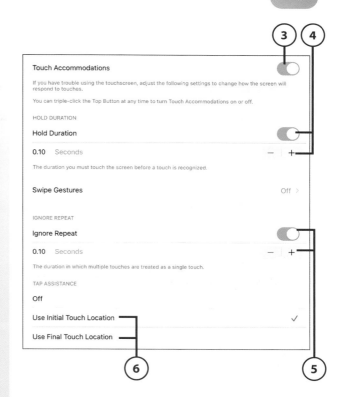

Enable AssistiveTouch

Your iPad includes an accessibility feature dubbed AssistiveTouch that overlays a group of large icons onscreen for common functions. With AssistiveTouch enabled, just tap the AssistiveTouch icon and then tap an icon for Notifications, Device, Control Center, Home, Siri, or Custom (any operation of your choice).

1 Go to the Settings screen, select the Accessibility tab, and then go to the Physical and Motor section and tap Touch.

2 Tap AssistiveTouch.

(3) Tap "on" the AssistiveTouch switch.

(4) The AssistiveTouch icon now appears at the bottom-right corner of the iPad screen. Tap this icon to display the AssistiveTouch Menu. Tap the action you want to initiate. (Tap anywhere outside the AssistiveTouch menu to close the panel.)

Use Siri

If you have trouble managing the gestures you need to operate your iPad, use voice commands instead. You can activate many accessibility features by talking to your iPad via the Siri personal digital assistant. Just say "Siri, turn on VoiceOver," for example, and that feature will be turned on. (Learn more in Chapter 7, "Controlling Your iPad–and More–with Siri.")

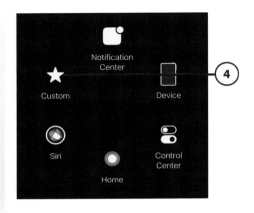

In this chapter, you learn how to connect your iPad to Wi-Fi networks and hotspots to access sites and information on the Internet. You also learn how to use the Safari web browser to surf the Web.

→ Connecting to a Wi-Fi Network
→ Using the Safari Web Browser
→ Making the Web More Readable

5

Connecting to the Internet and Browsing the Web

A lot of what you do with your iPad requires a connection to the Internet. Sending and receiving email, searching for information, video chatting with FaceTime, watching streaming videos, or listening to streaming music—all of these activities require an Internet connection.

You connect your iPad to the Internet via a Wi-Fi wireless connection. Your iPad has built-in Wi-Fi, so it's just a matter of connecting to a private Wi-Fi network (like you probably have at home) or a public Wi-Fi hotspot. After you're connected, you can do all that fun Internet-related stuff—as well as use those apps that also require an Internet connection.

Connecting to a Wi-Fi Network

To get the most out of your iPad, you need to connect it to the Internet via a wireless network. You can connect to your home Wi-Fi network or to any public Wi-Fi network or hotspot.

Connect to Your Home Wireless Network

Obviously, if you have a wireless network in your home, you'll want to connect your iPad to that. Once you've initially connected, you won't need to manually reconnect in the future; your iPad remembers the network settings and reconnects automatically.

To connect to your home network, you need to know the name of the network (sometimes called an SSID, for *service set identifier*) and the network password (sometimes called a *security key*).

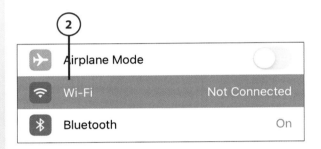

1. Tap the Settings icon to open the Settings screen.

2. In the left column, tap Wi-Fi. (If you're not yet connected, it should say "Not Connected.")

3. Make sure the Wi-Fi switch is in the "on" position. If it isn't, tap to turn it on.

4. You see a list of the wireless networks within range. Tap to select your home network. This displays the password panel.

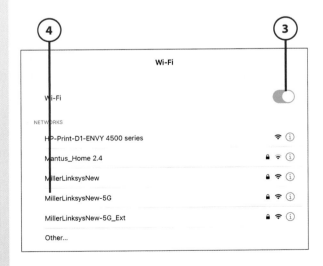

Public Versus Private Networks

Private wireless networks require a password to connect and are indicated with a lock icon in the network list. Public networks (sometimes called *open networks*) let you connect without a password and are not accompanied by the lock icon.

⑤ Use the onscreen keyboard to enter the password for your network.

⑥ Tap Join. You connect to the network and can access the Internet.

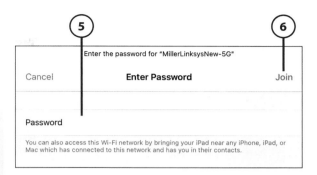

Current Connection

If you're currently connected to a wireless network, you see that network's name in the Wi-Fi field in the left column of the Settings page. You can connect to a different network if you like; see the "Change Networks" task later in this section.

Connect to a Public Wireless Hotspot

Most wireless networks and hotspots you find out in the real world are public networks—that is, you don't need a special password to connect. All you have to do is identify the network from the list and click to connect. (You'll know you're connected when you see the Wi-Fi icon in the status bar at the top of the screen.)

Some public networks, however, do require you to read through and agree to their terms of service after you connect but before you can use the network.

If this is the case, the site typically launches a connection page in your web browser. (In some instances, you may need to launch the Safari browser manually and try to access a web page; this will then launch the wireless connection page.) When you see the connection page from the wireless network or hotspot, you might need to check an "I read that" option or click a "connect" button.

For example, if you connect to the wireless networks at your local Starbucks or McDonald's, you need to manually log in from their respective log-in pages. In some instances, there may be a fee to use the wireless network, in which case you'll need to enter your credit card number to proceed.

Hotel Wi-Fi

If you're connecting to the wireless network in a hotel without free Wi-Fi, you typically enter your room number to add the charges to your bill. Given that some hotels charge $10 or more per night, you may want to shop around for hotels that offer free Wi-Fi.

(1) From the Settings screen, tap Wi-Fi in the left column.

(2) Make sure the Wi-Fi switch is in the "on" position. If it isn't, tap to turn it on.

(3) You see a list of the wireless networks and hotspots within range. Tap to select the desired network.

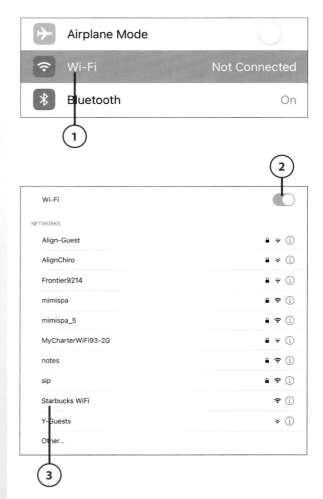

4 If the hotspot requires you to sign in, it should launch a log-in screen and display a sign-in page. (If this doesn't happen automatically, you might need to manually launch the Safari web browser and then enter any web page address to open the sign-in page.) Enter any necessary information, check to agree to any terms and conditions, and then click the "accept" or "sign in" button.

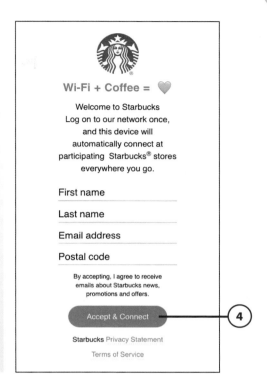

Wi-Fi + Coffee =

Welcome to Starbucks
Log on to our network once,
and this device will
automatically connect at
participating Starbucks® stores
everywhere you go.

First name

Last name

Email address

Postal code

By accepting, I agree to receive
emails about Starbucks news,
promotions and offers.

Accept & Connect

Starbucks Privacy Statement

Terms of Service

It's Not All Good

Public Wi-Fi

Public Wi-Fi is a popular and convenient way to connect to the Internet when you're out and about, but it's riskier than using a private Wi-Fi network like you have at home. Hackers can connect to public networks and hijack passwords and other sensitive information. For this reason, you should avoid entering your banking or credit card information, passwords for sensitive data, and other private information when browsing on a public Wi-Fi network. Save your online shopping and other private activities until you're at home.

Change Networks

Sometimes, especially out in public, you may have several different Wi-Fi networks available to you—and your iPad might automatically connect to one other than the one you wanted to connect to. If this is the case, it's easy enough to switch networks.

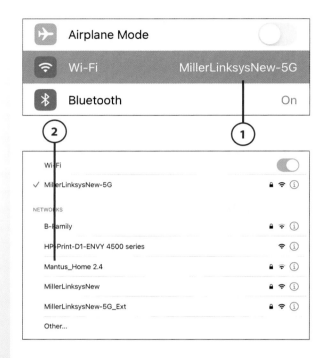

1. From the Settings screen, observe the Wi-Fi field in the left column. The currently connected network is displayed. If you don't want to connect to this network, tap the Wi-Fi field to display a list of other available networks.

2. Tap the network you want to connect to. Your iPad disconnects from the previous network and connects instead to the newly selected one.

Connect to a Network You've Previously Connected To

When you're in range of a wireless network or hotspot you've previously connected to, your iPad automatically connects to it again. You don't have to choose it from the list; the iPad automatically discovers the network and makes the connection.

If it's a private network, like your home network, your iPad remembers the password you previously used and enters it for you. If it's a public network or hotspot that requires logging in via your web browser, you might be asked to agree to terms and click the "sign in" button again. (Or you might not; some sites allow reconnection without having to sign back in.)

It's Not All Good

Connection Problems

Not all Internet connections are good. Sometimes the Wi-Fi network or hotspot you connect to has problems, which can keep your iPad from connecting to the Internet.

For example, the Wi-Fi hotspot I use at one of my local coffeehouses has a tendency to go missing every few hours. That is, I'll be connected one minute and the next minute find that I'm not connected—and that the hotspot itself is no longer visible on my iPad. Normally I wait a minute or two, the hotspot reappears, and my iPad reconnects. The best I can figure is that the coffeehouse's Wi-Fi router has rebooted for some reason, which kicks off everyone using it until it powers back on.

You also can run into similar connection problems with your home Wi-Fi network. In many cases, the problem corrects itself automatically within a few minutes. If the problem persists, try turning off your iPad's Wi-Fi and then turning it back on. This forces your iPad to establish a new connection to the wireless network or hotspot, which often fixes the problem.

Using the Safari Web Browser

After your iPad is connected to a Wi-Fi network or hotspot, you should have access (via that network) to the Internet. That's where all the fun stuff happens—including surfing the Web.

To surf the Web, you need an app called a *web browser*. You're probably familiar with using a web browser on your computer or phone, and it works similarly on your iPad.

The browser that Apple includes with your iPad is called Safari. It works like most other web browsers, and it's ready for use whenever you are.

Other Web Browsers

Safari is preinstalled on your iPad, but you can use other web browsers, if you like. The most popular non-Apple browser is Google Chrome, which is available for free from the App Store. Many people prefer Chrome to Safari, especially if they use Chrome on their computer or other devices. (All web browsers operate in generally the same way.)

Launch the Safari Browser

You launch Safari the same way you launch any other app on your iPad.

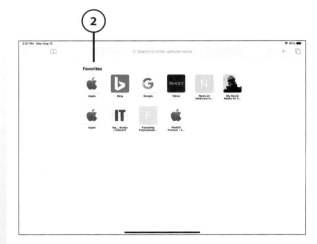

(1) Tap the Safari icon to open the Safari browser.

(2) The first time you launch Safari, it opens to a blank page, possibly with some website suggestions. If you've previously used the browser, it opens your last open page.

Enter a Web Address

To go directly to a given website or web page, you must enter the web address of that page into the Address box at the top of the browser screen.

(1) Tap within the Address box to display the onscreen keyboard; then enter the address of the web page you want to visit.

(2) As you type, Safari might suggest matching web pages. Tap any page to go to it. *Or…*

3 Finish entering the full address and then tap Go or press Enter on the keyboard to go to that page.

4 The web page appears onscreen. Swipe up to scroll down the page.

5 If a page has trouble loading, or if you want to refresh the content, you can reload the page by tapping the Reload button.

>>>*Go Further*
WEB ADDRESSES

A web address is called a Uniform Resource Locator, or URL. Technically, all URLs start with either http:// or (for secure sites) https://. You don't need to type this part of the URL, however; Safari assumes it and enters it automatically.

The main part of most web addresses starts with www followed by a dot, then the name of the website, then another dot and the domain identifier, such as com or org. As an example, my website is www.millerwriter.com. (You don't have to enter the "www."; Safari does this for you.)

Use Web Links

Pages on the Web are often connected via clickable (or, in the case of your iPad, tappable) links, called *web links*. A web link can be within a page's text (typically underlined or in a different color) or embedded in an image. Tap a link to go to the linked-to page.

(**1**) On the current web page, tap the web link.

(**2**) The linked-to page displays.

View Multiple Pages in Tabs

Safari, like most modern web browsers, lets you open more than one page at a time, using *tabs*. You can open different web pages in different tabs and easily switch between them.

(**1**) Tap the + icon to create a new tab.

(**2**) Enter the desired web address into the new tab's Address box.

(**3**) Tap a different tab to view that tab. *Or…*

(**4**) Tap the tabs icon to view thumbnails of all open tabs.

(**5**) Tap to switch to a different tab.

(**6**) Swipe a tab to the left to close it. *Or…*

(**7**) Tap the X on any tab to close it.

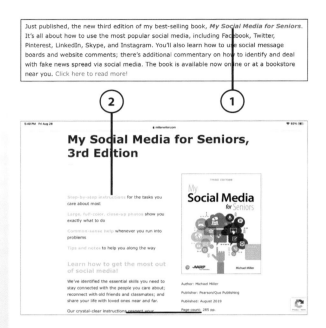

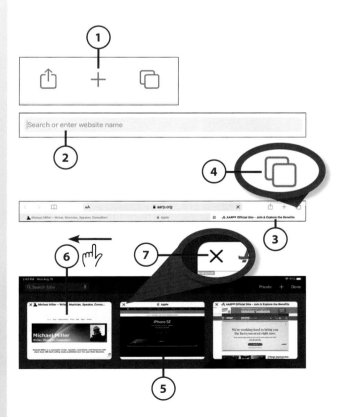

Bookmark Pages

You can save any page you visit by *bookmarking* that page. It's easy, then, to revisit those pages you've bookmarked.

(1) Navigate to the page you want to bookmark and then tap the Share icon. The Share panel displays.

(2) Tap Add Bookmark.

(3) Accept, edit, or enter the web page name in the Add Bookmark panel.

(4) By default, bookmarks are saved in the Favorites folder. To save this bookmark to a different folder, tap the Location field and choose a different location.

(5) Tap Save. The page is now saved in your bookmarks.

Favorites

Safari offers a special category of bookmarks, called Favorites. You save a page to your Favorites list by tapping Add to Favorites in the Share panel. All Favorites can be accessed by selecting Favorites in the Bookmark panel; Favorites are also displayed when you tap + to open a new empty tab.

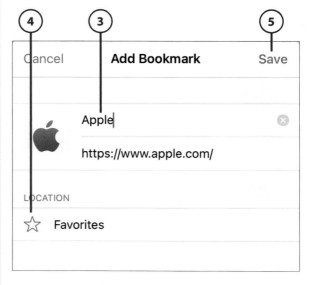

6. Tap the Bookmark icon to open the Bookmark panel so you can view and revisit bookmarked pages.

7. Make sure the Bookmarks tab is selected.

8. Some bookmarks are organized into folders. Tap a folder name to view all bookmarks in that folder.

9. Tap the name of a bookmark to open the corresponding web page.

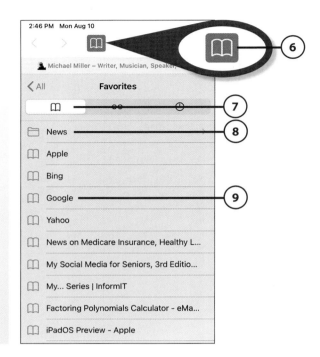

>>>Go Further
ADD TO HOME SCREEN

If you find yourself visiting a given web page with regularity, you might want to create a shortcut to it on your iPad's Home screen. When you add an icon for a page to the Home screen, tapping that icon launches Safari and opens that page.

To add an icon for a given web page, navigate to that page, tap the Share icon, and then tap Add to Home Screen. When the Add to Home panel appears, tap Add.

Revisit Past Pages

How easy is it to return to a web page you've previously viewed? Safari keeps track of all your web browsing history, and revisiting a page is as easy as tapping it in the Bookmark panel.

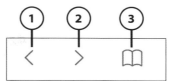

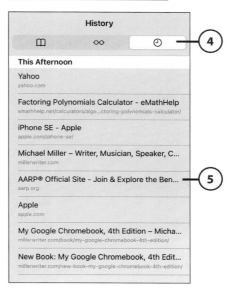

1. View the page you just visited by tapping the back (left arrow) icon at the top-left corner of the browser.

2. Return to the next page by tapping the forward (right arrow) icon.

3. Tap the Bookmark icon to open the Bookmark panel.

4. Tap History to view pages you've visited; the newest is listed first.

5. Tap a page to go to that page.

Browse the Web in Private

There are some web pages you might want to browse in private and not let others know you've seen. When you want to browse anonymously, use Safari's Private mode. Pages you visit while in Private mode are not stored to your history and are not otherwise tracked on your device or elsewhere.

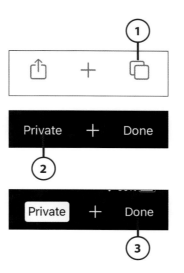

1. From within Safari, tap the Tabs icon.

2. Tap Private.

3. Tap Done. You are now in private browsing mode; the top of the page background is dark.

(4) Use Safari as normal. None of your activity will be tracked.

(5) Tap the Tabs icon to exit private browsing mode.

(6) Tap Private. You see any tabs you had open before entering private browsing mode.

(7) Tap Done.

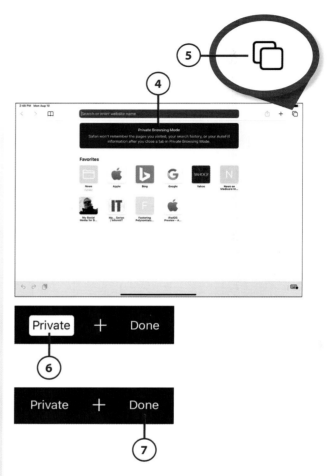

Configure Safari Settings

Safari is configured perfectly for most users by default, but there are a number of settings you can change to better personalize your Safari browsing experience. You access all these settings from your iPad's Settings screen.

(1) Tap the Settings icon to open the Settings screen.

(2) Scroll down the left column to the apps section and tap Safari.

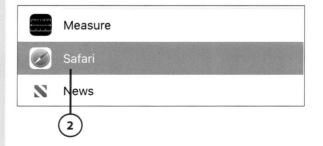

(3) Tap to edit any setting.

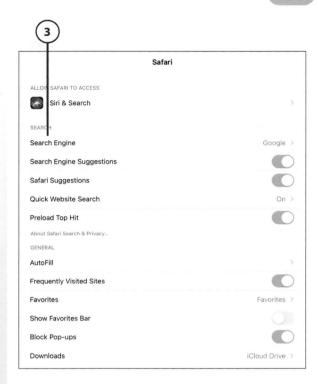

Clear History, Cookies, and Browsing Data

From time to time, you may want or need to clear accumulated browsing data—information about the sites you've visited—from the Safari browser. You do this from the Settings screen.

Clear the Cache

Web pages you've visited are stored in what is called a *cache*, which helps Safari load those pages faster if you revisit them. If Safari has trouble displaying or accessing a web page, you may be able to rectify the issue by clearing the browser's cache. You do this by using the Clear History and Website Data function.

(1) From the Settings screen, scroll down the left column to the apps section and tap Safari.

(2) Tap Clear History and Website Data.

(3) When the pop-up panel appears, tap Clear.

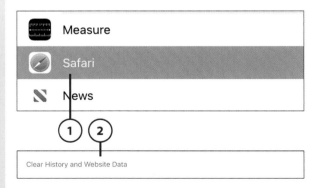

>>>*Go Further*

BLOCK ONLINE ADS

If you're like me, you quickly tire of all the online ads displayed on various web pages. (There are even ads on Facebook and many web-based email services!) Fortunately, you can block many, if not all, of these ads by installing an ad blocker extension in the Safari web browser. With an ad blocker installed, ads simply don't display on a web page—you see a blank space where the ad is supposed to be. Ad blockers don't hurt your browser or your iPad and make browsing the Web more pleasant.

There are a variety of different ad blockers available in the App Store. Some are free; some aren't. The most popular ones include Ad-Blocker Pro, AdBlock, and Norton Ad Blocker.

You may have to manually enable any ad blocker you install. From your iPad's Settings screen, tap Safari in the left column and then tap Content Blockers. (This option is not available until you install an ad blocker.) You can then enable or disable any ad blocker you've installed.

Making the Web More Readable

In years past, Safari displayed the mobile versions of many web pages—the same versions displayed on smartphones. This meant you may have seen very small type or columns that wrapped too far off the side of a page. Beginning with iPadOS 13 and continuing into iPadOS 14, Safari always displays the non-mobile versions of web pages, which should make everything more readable.

The emphasis here is on *should*. There are still many web pages out there that use too-small type, which makes them difficult to read on a tablet device. Fortunately, there are ways to make small type on web pages a little larger—and web pages more readable, in general.

Zoom in to a Page

If you find a particular web page difficult to read, the first thing you can do is try to zoom in to the text—that is, make the text larger onscreen.

1. Place two fingers together on the screen and then expand them. As you move your fingers apart, you zoom in to that portion of the page.

2. To zoom back out, pinch your two fingers back together.

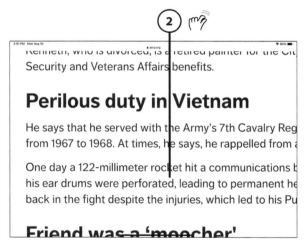

Read Articles with Safari Reader

Safari Reader is a special reading mode that makes some web pages easier to read by removing ads, images, videos, and other extraneous elements. Not all web pages can be viewed with Safari Reader, but those that can are a lot easier to read than normal cluttered web pages. (Safari Reader is especially useful for reading articles on web news sites.)

1. Tap the AA icon in the address bar.

2. Tap Show Reader View. (This option is available only if a page can be viewed with Safari Reader.)

3. You see the page in Reader View without distracting graphics and ads.

4. To return to normal web page view, tap the text icon again.

5. Tap Hide Reader View.

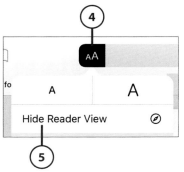

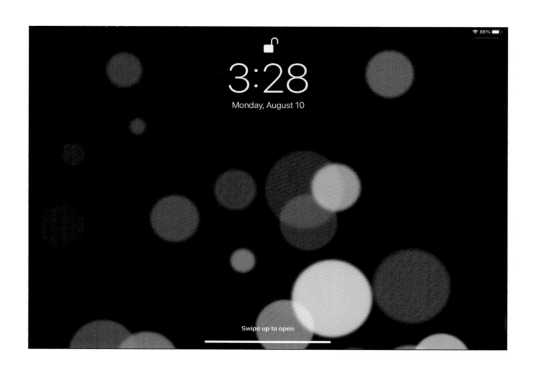

In this chapter, you learn how to keep your
iPad safe from unwanted users.

6

Keeping Your iPad Safe and Secure

Even the largest iPad is a relatively small portable device. That means that it could be relatively easy for someone to walk away with it while you're not looking. If someone takes your iPad—accidentally or on purpose—do you really want them digging around in everything you have stored inside?

Fortunately, there are ways to keep the information on your iPad safe from prying eyes. And there are lots of things you can do to minimize the risk of data theft if your iPad is lost or stolen.

In addition, there are things you can do to keep your information and identity safe when you use your iPad to go online. The Internet can be a dangerous place, so it makes sense to be as safe as possible when you're connecting with your iPad.

Creating a Safer Lock Screen

When it comes to keeping unwanted users away from the information stored on your iPad, your first and best line of defense is the device's Lock screen. If you play your cards right, strangers simply won't be able to get past the Lock screen to see anything else on your iPad.

Set a Simple Passcode

By default, you unlock your iPad by pressing the Home button or, on a model with Face ID, looking at and swiping up on the screen. That's convenient but far from secure; anybody picking up your device can unlock and see what you've stored inside.

A better approach is to establish a passcode (like a personal identification number, or PIN) that you use to unlock your phone. Your passcode should consist of six numbers and be fairly random—or as random as you can remember.

When you've enabled passcode operation, you're prompted to enter your passcode after you've restarted your iPad or woken it from sleep. (Unless you've enabled Face ID on selected models, that is.)

Changing Your Unlock Method

If you've already set a passcode, Touch ID print, or Face ID scan, you can change it by following the steps here. When you tap Touch ID & Passcode or Face ID & Passcode from the Settings screen, you're prompted to enter your passcode before you can change it.

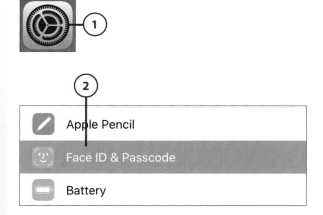

1. Tap the Settings icon to open the Settings screen.

2. Tap Touch ID & Passcode or Face ID & Passcode in the left column. (Which option you see depends on whether your iPad has a Home button.)

3 If you've already set a passcode, you're prompted to enter it for security reasons.

4 Scroll down and tap Turn Passcode On to display the Set Passcode panel.

5 Enter your desired six-digit passcode.

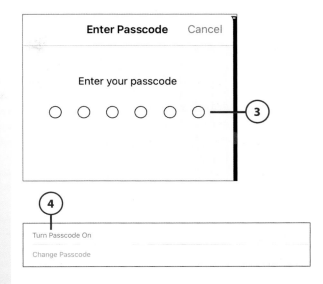

Creating a Passcode

Your passcode should be a series of digits you can remember but not something easily guessed. For example, you shouldn't use your birthdate as your passcode, nor should you set a common series of digits, such as 123456.

6 When prompted, reenter your six-digit passcode. Your passcode is enabled; you'll need to use it to unlock your device in the future—so write it down somewhere only you can find it!

Data Encryption

Creating a passcode also turns on your iPad's data encryption. This uses your passcode as a key to encrypt Mail messages and all attachments stored on your device.

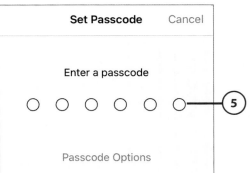

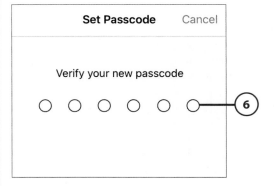

Use Fingerprint Recognition with Touch ID

If your iPad has a Home button, you can employ even stronger security by using your fingerprint to unlock your device. This is called Touch ID, and you can also use it to make purchases in the iTunes Store and App Store, provide debit and credit card payment information, enter billing and shipping addresses, and provide contact info when making in-app purchases on apps that offer Apple Pay as a payment method.

To unlock your phone using Touch ID, all you have to do is press your thumb or finger (whichever fingerprint you registered) on the Home button, which doubles as the Touch ID sensor. When the iPad recognizes your fingerprint, it unlocks.

Here's how to configure Touch ID.

(1) From the Settings screen, tap Touch ID & Passcode in the left column. (If asked to enter your passcode to proceed, do so.)

(2) Tap Add a Fingerprint.

(3) Place your finger or thumb on the Home button.

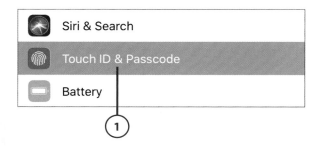

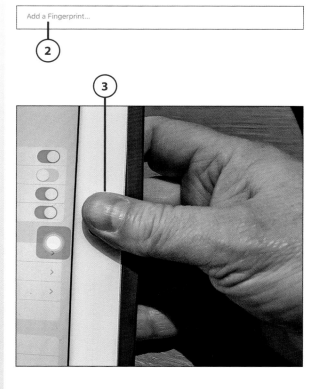

(4) Follow the onscreen instructions to lift and place your finger as many times as necessary to register your fingerprint.

(5) When the process is complete, tap Continue. Your device is ready to use with Touch ID enabled.

Adding Other Fingers

You can set up multiple fingers for your Touch ID. For example, you could set your forefinger and your thumb if you like to use both. Or if you have multiple people using your iPad, you can set up other people's fingers for Touch ID login, too.

iPad Air

If you have the fourth-generation iPad Air, it doesn't have a Home button but still uses Touch ID via the Top button. Follow these same instructions to configure Touch ID, but touch the Top button where the steps say to touch the Home button.

(4)

Place Your Finger

Lift and rest your finger on the Home button repeatedly.

Complete

Touch ID is ready. Your print can be used for unlocking your iPad.

Continue

(5)

Use Face Recognition with Face ID

If you have an iPad Pro model, you can unlock your device just by smiling at it. Your iPad uses its front-facing camera to scan your face and employ Face ID face recognition technology to verify your identity and unlock your device.

To unlock your iPad with Face ID, tap the screen to wake up the device and then stare at the screen. (Make sure you don't cover the camera with a finger.) When the lock icon changes from closed to open, swipe up from the bottom of the screen, and it's ready to go.

Here's how you configure Face ID.

1. From the Settings screen, tap Face ID & Passcode in the left column. (If asked to enter your passcode to proceed, do so.)

2. Tap Set Up Face ID. (Or, to rescan your face, tap Reset Face ID and then Set Up Face ID.)

3. Hold your iPad in portrait mode, and then tap Get Started.

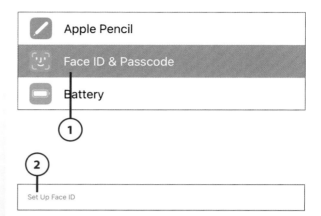

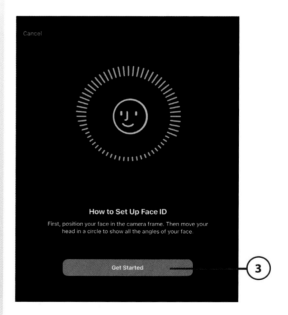

4 Position your head inside the frame and slowly move your head from side to side.

5 When this first scan is complete, tap Continue.

6 Repeat step 4 to rescan your face.

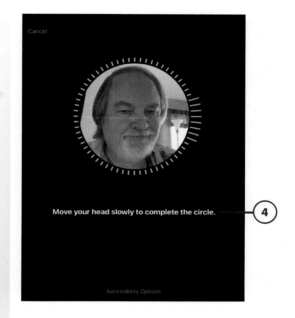

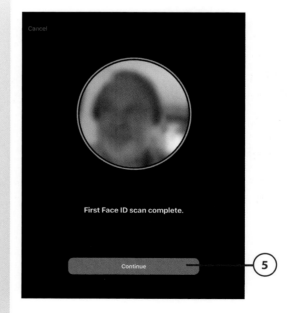

7. When the second scan is complete, tap Done.

Erase Data After Too Many

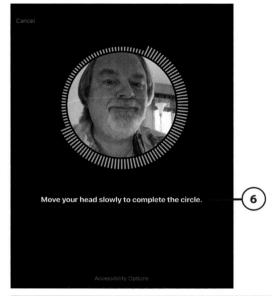

Move your head slowly to complete the circle. — 6

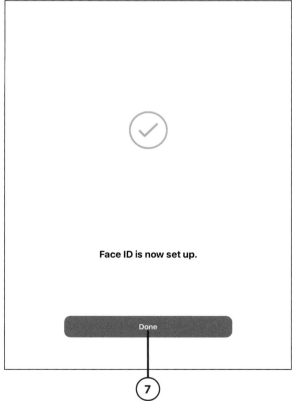

Face ID is now set up.

Done — 7

Unlock Attempts

As the ultimate "fail-safe" procedure, you can configure your iPad to erase all its stored data after 10 failed passcode attempts. This ensures that none of your data will fall into the wrong hands, especially if your iPad is stolen.

(1) From the Settings screen, tap Touch ID & Passcode or Face ID & Passcode in the left column. (If asked to enter your passcode to proceed, do so.)

(2) Scroll down and tap "on" the Erase Data switch.

(3) Tap Enable in the confirmation box.

Don't Forget Your Passcode

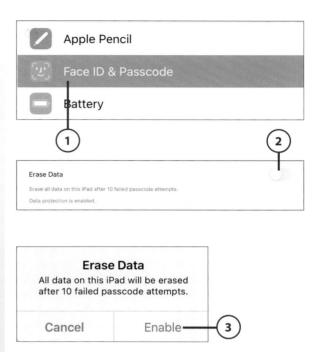

It's Not All Good

If you forget your passcode and enter more than 10 passcode attempts, you need to restore your iPad to use it again. Restoring your iPad to factory condition also deletes all the apps and data stored on the device—although you can restore your data from a backup, as discussed in Chapter 24, "Managing Files on Your iPad and in the Cloud." So don't forget to back up your data, and write down your passcode in a safe place.

Reducing the Risk of Theft

As noted, your iPad is a relatively small device that's usually not plugged in to anything, which means it can be easily heisted by those with impure hearts. (It's also easy to lose or leave behind, but that's another issue.) Given the attractiveness of the iPad to would-be thieves, how do you keep it away from sticky-fingered strangers?

Keep Your iPad Safe

Your iPad is a particularly attractive target for thieves. Not only is it small and easy to steal, it's also a high-value item. All Apple products are considered luxury items, and thieves like items that can command a high price when sold or pawned.

Knowing this, you need to take extra care when you take your iPad out in public. Here are some precautions you've probably already considered but are still important to keep in mind:

- Don't show it off. You may be very proud of your shiny new iPad, but don't make a big deal out of it. When you're in public, don't make others aware of what you have. Treat it as low key as you can.

- Keep it hidden—especially when you're not around. An expensive iPad is not the sort of thing you want to leave sitting on the table at a coffee shop when you go off to use the restroom. It's also not a good thing to leave visible on the front seat or dashboard of your car when you park. When you're not using it, put it away—in your purse or backpack or briefcase. Or just carry it around with you.

- Be aware of your surroundings. A near-empty coffeehouse with a table close to the counter is safer than a crowded fast-food restaurant filled with suspicious-looking characters. If the surroundings look unsafe, keep a firmer grasp on your iPad—or don't bring it out at all.

- Be cautious around strangers. Yes, there will be earnest individuals who see you using your iPad and want to ask you all sorts of questions about it. There also are smooth-talking heisters who get you to show them your iPad purely as a ruse to grab it out of your hands. Beware the latter.

- Keep it safe—even at home. You might think it would be okay to leave your iPad sitting out on the coffee table at your house or apartment, but that's just tempting fate. If your house is broken into, small expensive items sitting in plain sight are the first things that thieves grab. And if your house has a lot of foot traffic (think friends of your kids or grandkids and tradespeople), a loose iPad on the table is pretty attractive for anyone with money or addiction problems. Even at home, keep your iPad in a safe and private place.

In other words, be smart about how you use and store your iPad, and you can remove a lot of the risk involved.

Track a Lost or Stolen iPad

Your iPad includes an app, called Find My, which lets you track any lost or stolen iPad or iPhone, even if it's not connected to a Wi-Fi network (or, for cellular models, to either a Wi-Fi or cellular network). A lost device can communicate with other nearby Apple devices via Bluetooth to relay its location back to you.

To use Find My, you must have another Apple device in addition to the iPad you want to track. This can be another iPad, an iPhone, or even a Mac computer.

If your lose your iPad, or if someone steals it, your device periodically emits an encoded signal via Bluetooth. (As long as it has power, that is.) Any nearby Apple device running iPadOS or iOS, or macOS Catalina, will pick up that signal and transmit it back to Apple. You can then access that signal from the Find My app on any other Apple device just by signing in to your Apple account. (Only devices signed in to your account can decode the signal, so your privacy is maintained.)

Here's how to use the Find My app to find your device. Make sure you're signed in to your Apple account on the other device, and then follow these steps.

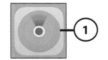

① Tap the Find My icon to open the Find My app.

② Tap the Devices tab in the panel at the lower left.

3 All your devices are displayed on the full-screen map; zoom in to the map if necessary. Tap to select the lost or stolen device.

4 Expand the new panel at the bottom left. If you see the iPad on the map but don't know how to get there, tap Directions to display turn-by-turn directions to that location.

5 Tap Play Sound to play an alert sound on the missing iPad for two minutes. (This works even if the device has been muted.)

6 Tap "on" the Notifications switch to receive a notification when the missing iPad is found.

7 Tap Activate in the Mark as Lost section to immediately lock the missing iPad with a passcode and display a custom message onscreen, along with your contact number. (Lost mode automatically enables Location Services, which is used to track your device's location.)

8 Scroll down to the bottom of the panel and tap Erase This Device to delete all the data and media on your iPad and restore it to the original factory settings. Employ this option as a last resort to keep anyone from accessing your personal data or using your iPad to make unauthorized payments with Apple Pay.

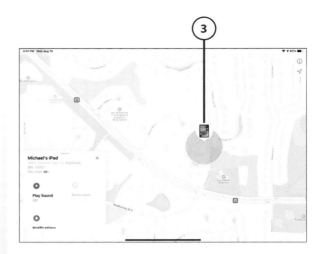

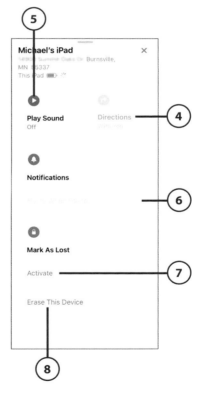

>>>*Go Further*

SHARE YOUR LOCATION WITH OTHERS

You can also use Find My to share your location with others. From within the Find My app, tap the People tab and tap Start Sharing Location. You can then enter the names of people in your contacts list so they can see your location in their Find My apps.

Other people with Apple devices also can use Find My to share their location with you, so you'll see them in your Find My app. It's a great way to keep track of children and other family members.

Staying Safe Online

Much of what you do on your iPad is done over the Internet, via either the Safari web browser or apps that use online resources. Unfortunately, the Internet can be a dangerous place. There are certain predators who target older users online, and for good reasons. Many older individuals are more trusting than younger users, and many are also less tech-savvy. In addition, many older people often have large nest eggs that can be particularly attractive to online predators—and some older individuals can be ashamed to report being taking advantage of.

We all need to learn how to protect ourselves when we're online. You need to be able to identify and avoid the most common online threats and scams.

Protect Against Identity Theft

Online predators want your personal information—your real name, address, online usernames and passwords, bank account numbers, and the like. It's called *identity theft*, and it's a way for a con artist to impersonate you—both online and in the real world. If your personal data falls into the hands of identity thieves, they can use it to hack into your online accounts, make unauthorized charges on your credit card, and maybe even drain your bank account.

There are many ways for criminals to obtain your personal information. Almost all involve tricking you, in some way or another, into providing this information of your own free will. Once you know what to look for, you're better able to protect yourself.

One of the most common techniques used by identity thieves is called *phishing*. It's called that because the other party is "fishing" for your personal information, typically via fake email messages, text messages, and websites.

A phishing scam typically starts with a phony email or text message that appears to be from a legitimate source, such as your bank, the postal service, PayPal, or other official institution. Some phishing messages purport to be from someone you personally know, such as your manager at work or your pastor at church.

Most phishing messages purport to contain important information that you can see if you tap the enclosed link. If you tap the link, you're taken to a fake website masquerading as the real site. You're encouraged to enter your personal information into the forms on this fake web page; when you do so, your information is sent to the scammer, who can now steal your identity.

How can you avoid falling victim to a phishing scam? There are several things you can do:

- Look at the sender's email address. Tap it to view more detail. Most phishing emails come from an address different from the one indicated by the (fake) sender.

- Look for poor grammar and misspellings. Many phishing schemes come from outside the United States by scammers who don't speak English as their first language. As such, you're likely to find questionable phrasing and unprofessional text—not what you'd expect from your bank or other professional institution.

- If you receive an unexpected message, no matter the apparent source, do *not* tap any of the links included. If you think there's a legitimate issue from a given website, go to that site manually in Safari and access your account from there.

- Some phishing messages include attached files that you are urged to open to display a document or image. Do *not* open any of these attachments; they might contain malware that can steal personal information from your device. (Although malware isn't as much of a threat to iPads as it is to personal computers, you still want to avoid opening unexpected email attachments.)

Malware Threats

If you're a long-time computer user, you're probably familiar with the threat posed by computer viruses, spyware, and other malicious software (malware). Fortunately, due to technological safeguards built into the iPadOS operating system, the malware risk for iPad users is extremely low. So don't worry too much about malware on your iPad—but still remain vigilant, nonetheless.

Keep Your Private Information Private

Identity theft can happen any time you make private information public. This has become a special issue on social networks, such as Facebook, where users tend to forget that most everything they post is publicly visible—and often share information about birthdays, schools attended, pets' names, and even phone numbers.

None of this might sound dangerous until you realize that all of these items are the type of personal information many companies use for the "secret questions" their websites use to reset users' passwords. A fraudster armed with this publicly visible information could log on to your account on a banking website, for example, reset your password (to a new one he provides), and thus gain access to your banking accounts.

The solution to this problem, of course, is to enter as little personal information as possible when you're online. For example, you don't need to—and shouldn't—include your street address or phone number in a comment or reply to an online news article. Don't give the bad guys anything they can use against you!

Protect Against Online Fraud

Identity theft isn't the only kind of online fraud you might encounter. Con artists are especially creative in concocting schemes that can defraud unsuspecting victims of thousands of dollars.

Many of these scams start with an email or social media message that promises something for nothing. Maybe the message tells you that you've won a lottery or you are asked to help someone in a foreign country deposit funds in a U.S. bank account. You might even receive requests from people purporting to be far-off relatives who need some cash to bail them out of some sort of trouble.

The common factor in these scams is that you're eventually asked to send money (typically via wire transfer) or provide your bank account information—with which the scammers can drain your money faster than you can imagine. The damage can be considerable.

Gift Card Scams

Another popular scam consists of an email message that looks like it comes from someone you trust—such as the pastor of your church or an executive at the company you work for. The message asks you to go to a specific store and purchase some gift cards for that person. If you do so and provide the sender with the gift card numbers, you're out that cash.

Most online fraud is easily detectible by the simple fact that it arrives out of the blue and seems too good to be true. So if you get an unsolicited offer that promises great riches, you know to tap Delete—pronto.

Savvy Internet users train themselves to recognize these scam messages. That's because most scam messages come from complete strangers and often don't even address you by name. Most of these messages have spelling and grammatical errors because scammers frequently operate from foreign countries and do not use English as their first language. Con artists know their trade well, and even the smartest, most educated people can get scammed. Knowing what to look for is key.

If you receive a message that you think is a scam, delete it. In fact, it's a good idea to ignore all unsolicited messages of any type. No stranger will send you a legitimate offer via email or Facebook; it just doesn't happen.

>>>*Go Further*

WHAT TO DO IF YOU'VE BEEN SCAMMED

What should you do if you think you've been the victim of an online fraud? There are a few steps you can take to minimize the damage:

- If the fraud involved transmittal of your credit card information, contact your credit card company to put a halt to all unauthorized payments—and to limit your liability to the first $50.

- If you think your bank accounts have been compromised, contact your bank to put a freeze on your checking and savings accounts—and to open new accounts, if necessary.

- Contact one of the three major credit reporting bureaus to see if stolen personal information has been used to open new credit accounts—or max out your existing accounts.

- Contact your local law enforcement authorities—fraud is illegal, and it should be reported as a crime.

- Report the issue to AARP's Fraud Watch Network, at 877-908-3360.

Above all, don't provide any additional information or money to the scammers. As soon as you suspect fraud, halt all contact and cut off all access to your bank and credit card accounts. Sometimes the best you can hope for is to minimize your losses.

Shop Safely

Many people use their iPads to shop online. It's convenient, and you don't have to bother with driving to the store and dealing with all those crowds. It's been critical during the COVID-19 pandemic.

Despite the huge upsurge in online shopping, many users are still reticent to provide their credit card information over the Internet. It is possible, after all, for shady sellers to take your money and not deliver the goods, or even for high-tech thieves to intercept your credit card information over the Internet—or from a public Wi-Fi hotspot.

All that said, online shopping remains immensely popular and is generally quite safe. You can minimize your risk when online shopping by following this advice:

- Don't shop when using public Wi Fi. Although you can shop over any wireless connection, public connections (like the kind you find at coffeehouses and restaurants) aren't secure. It's possible for individuals with the right equipment to intercept public wireless signals, and thus skim your credit card and other personal information. While this sort of data theft doesn't happen often, it's better to make your online purchases over a safer private connection.

- Shop only at secure websites. Whatever you're shopping for online, make sure you're using a website that offers secure connections. A secure web address starts with https:, not the normal http:. A secure website encrypts the data you send to it, so even if it is intercepted by a third party, that party can't read it. Most major online retailers have secure sites.

- Don't leave your credit card number on file with online retailers. As tempting as it is to have your favorite retailer store your credit card info for future purchases, that also means the retailer has a copy of it on its servers—and anyone breaking into its servers can then steal your information. Instead, enter your credit card number fresh with each new purchase; it's just safer.

If you shop at major online retailers and follow the advice presented in this chapter, you're probably going to be safe. Same thing if you buy from sellers on eBay or Etsy; those sites have their own robust security mechanisms in place. If it's a retailer you haven't heard of before, check it out by looking it up on the Better Business Bureau or doing a Google search and reading online reviews; if the reviews trend toward the negative, shop elsewhere. And always, always shop from a merchant that offers a toll-free number for customer support, just in case something goes wrong.

Do all of these things and shopping with your iPad will be not only convenient but also safe.

Fraud Watch Network

AARP's Fraud Watch Network is a free and valuable source of information about online scams and fraud. Check it out at www.aarp.org/fraudwatchnetwork.

Controlling Your iPad—and More—with Siri

Siri isn't a person; it's a thing, sort of. To be precise, Siri is a piece of software that functions as a virtual personal assistant on your iPad and other Apple devices. It's a way of both operating your iPad using voice commands and searching the Internet for relevant information.

Siri understands natural speech, so all you have to do is talk to your iPad using plain-English questions and commands. Siri knows what you're asking and responds accordingly. You can use Siri not just to ask questions and find information, but also to control the operations of your iPad.

Getting to Know Siri, Your iPad's Virtual Personal Assistant

Siri functions as a high-tech personal assistant on your iPad. In reality, Siri is nothing more than an app on your device, albeit one that is highly functional and very high tech. The Siri app essentially serves as a voice-controlled interface between you and various operations on your iPad, enabling you to perform many operations via voice commands.

You can use Siri to initiate most operations on your iPad. Use Siri to compose and send messages, schedule events, watch videos, and listen to music, as well as launch apps, turn on Do Not Disturb mode, and raise and lower your iPad's volume levels.

Siri can also keep you updated on current sports scores, news events, and weather conditions. You can even ask Siri factual questions, such as, "How many miles is it to the moon?" or "How much is five times seven?"

Configure Siri

There are several things you can change to personalize your own experience with Siri. You can change Siri's voice, the language she speaks, and even whether you can summon Siri merely by saying, "Hey Siri!"

1. From the Dock or Home screen, tap the Settings icon to open the Settings screen.

2. In the left column, tap Siri & Search.

3. To summon Siri without first tapping the Home button, tap "on" the Listen for "Hey Siri!" switch. You're prompted to set up Siri to recognize your voice; follow the onscreen instructions. (You may have set this up when you first turned on and configured your iPad; if so, this switch is already "on" and you don't have to do anything further.)

4. On iPads with a Home button, you can summon Siri by pressing and holding the Home button. This is enabled by default. To disable Home button activation, tap "off" the Press Home for Siri switch.

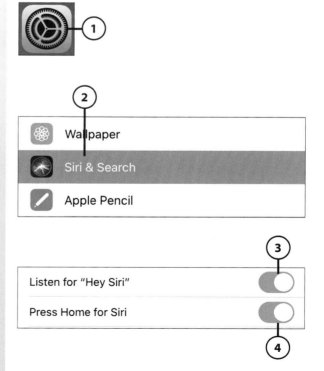

5 On iPads without a Home button, you can summon Siri by pressing and holding the On/Off (top) button. This is also enabled by default. To disable On/Off button activation, tap "Off" the Press Top Button for Siri switch.

6 By default, you can summon Siri from any screen, including the Lock screen (before you unlock it). If you don't want Siri available to your locked device, tap "off" the Allow Siri When Locked switch.

7 Tap Language and then make a selection to change the language that Siri speaks.

8 To change Siri's speaking voice, tap Siri Voice and make a new selection. Choose from American, Australian, British, Indian, Irish, or South African accents, in either Male or Female (default) voices.

9 By default, Siri provides voice and onscreen feedback. You can change this so that Siri only speaks when spoken to (and provides onscreen feedback otherwise). Tap Siri Responses and then tap Only with "Hey Siri."

10 Siri can operate more effectively if it knows more about you. Tap My Information and select your name from the resulting contacts list.

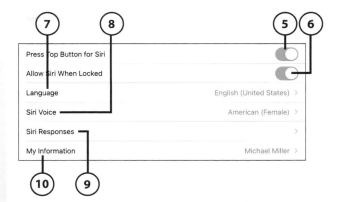

11 By default, Siri offers suggestions when you use your iPad's Search function. If you find this too intrusive, tap "off" the Suggestions While Searching switch.

12 Siri offers suggestions for things for you to do, based on your past activities, displayed on your iPad's Lock screen. This might include suggesting that you order your favorite drink from a coffee shop in the morning or play your favorite playlist in the afternoon. To turn off this feature, tap "off" the Suggestions on Lock Screen switch.

13 By default, Siri offers suggestions on the Home screen. If you'd rather not see these suggestions, tap "off" the Suggestions on Home Screen switch.

14 Also by default, Siri offers suggestions when you're sharing with other devices. To disable this, tap "off" the Suggestions when Sharing switch.

15 Siri can function within individual apps. To change how Siri works within a given app, tap the name of that app and then tap "on" or "off" any of the available options. These include Show Siri Suggestions in App, Learn from This App, Suggest App, Show App in Search, and Suggest Shortcuts for App.

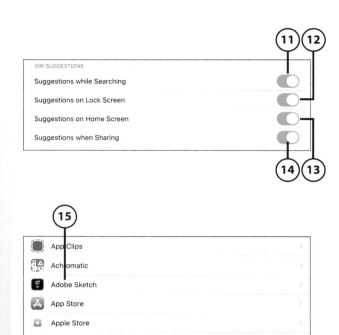

Summon Siri

There are four ways to summon Siri to do your bidding, depending on the iPad you have and how you've configured it.

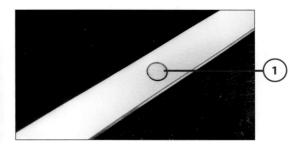

(1) On an iPad with a Home button, press and hold the iPad's Home button. *Or…*

(2) On an iPad without a Home button, such as the iPad Pro, press and hold the On/Off button. *Or…*

(3) If you have earphones with a microphone, hold down the control or mute button on your earphones. *Or…*

(4) If you've enabled "Hey Siri!" functionality, simply speak the words "Hey Siri!" into the device. (Not shown.)

(5) You'll see the Siri globe at the bottom-right corner of the screen. Speak your question or command.

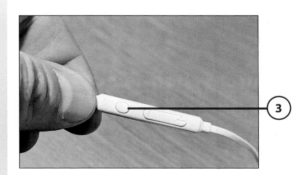

(6) Siri speaks her response and displays relevant information onscreen. Swipe right on this content box to remove it from the screen.

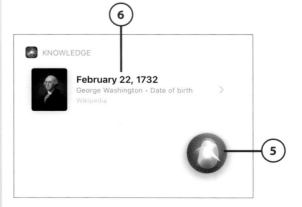

KNOWLEDGE

February 22, 1732
George Washington · Date of birth
Wikipedia

Louder or Softer

To adjust the volume level of Siri's voice, use the volume up and down buttons on the side of your iPad.

Using Siri to Control Your iPad and Apps

Instead of tapping onscreen icons or controls, you can use Siri's voice commands to perform many operations on your iPad. This makes Siri essential for using the iPad for those with accessibility issues—and just more convenient for the rest of us.

Connection Needed

To use Siri, your iPad must be connected to the Internet.

Launch Apps

The first thing you can do with Siri is launch apps on your iPad. To do so, say something like

Launch [app name].

For example, to launch the Facebook app, say

Launch Facebook.

Control System Functions

You can use Siri to enable, disable, and control various system functions on your iPad. Say the following:

Turn Wi-Fi on.
Turn Bluetooth off.
Increase the brightness.

Send and Receive Messages

Siri is useful if you want to send or receive messages via the Messages or Mail apps. To read messages, say the following:

Read my messages.
Read messages from Bob.
Read my last message from Sally.

To reply to a message, say the following, and then say "Send:"

Reply I'll see you then. Send.
Reply okay. Send.

To create a new message, say any of the following, and then say "Send:"

New email to John Jones. (Then dictate your message.)
Send a message to Oliver saying I'm free tomorrow.
Email Clark and say I received the pictures, thanks.

You can also use Siri to initiate FaceTime conversations. Say something like

FaceTime Hayley.
End conversation.

Manage Meetings, Events, and Reminders

If you use your iPad to manage your meetings, events, and reminders, Siri definitely can come in handy.

Here are a few commands to try regarding meetings and appointments:

What's on my calendar tomorrow?
Do I have a meeting at noon?
Where is my 10 o'clock meeting?
Set up a meeting with Hal at 2.
Cancel my 11 o'clock appointment.
Reschedule my meeting with Dinah to next Friday at 10.

You can even create complex events, complete with other attendees. Just tell Siri as much as possible about the event, like this:

Make an event for 9:45 tomorrow called Neighborhood Association with Randy Jones, Don Emory, Betsy Griffin, and Samuel Hancock.

Keep It Simple

The more complex your instructions, the higher the possibility that Siri won't get it right. (Or that you'll say it wrong. That happens.)

Then there are reminders, of which Siri can be quite helpful. Say something like

Remind me to pay the electric bill by the 15th.
Remind me to stop at the drug store when I leave here.
Remind me to buy milk.
Remind me to feed the fish when I get home.

You can also use Siri to create notes. Say something like

Create a new note or *Make a note named Parts List,* and then dictate your note.
Find my note about plumbers.

You can even use Siri to set alarms and timers. Say something like

Set an alarm for 6 a.m.
Set an alarm for three hours from now.
Turn off all alarms.
Set a timer for 15 minutes.

Take and View Pictures

If you use the camera in your iPad, you can control it via Siri voice commands.
Say something like

Open camera.
Take a picture.
Take a selfie.

To view pictures you've taken, say something like

Show me photos from October.
Show me photos I took last year.
Show me photos of Dave.
Show me photos of Florida.
Show me photos from my Family album.

Listen to Music

If you use your iPad's Music app to listen to music, you can control what you
listen to via Siri's voice commands. Try the following:

Play music.
Pause music.
Stop.
Next song.
Previous song.

To play specific music, say something like

Play James Taylor.
Play "Proud Mary."
Play Abbey Road.
Play the playlist My Favorites.
Play the radio station Classic Rock.

You can even shuffle the music you call up:

Play Classic Soul shuffled.

To find out more about the currently playing song, say something like

What's playing?
Who sings this song?
Who is this song by?

>>>*Go Further*

SMART HOME CONTROL

If you use Apple's Home app to control HomeKit-enabled devices in your home, you can use Siri to control those devices via voice commands. For example, you can tell Siri to do the following:

Turn off the lights.
Turn down the dining room lights.
Dim the lights.
Set brightness to 50 percent.
Set the temperature to 72 degrees.
Turn on the coffeemaker.

After you get a device set up, experiment with different commands to see just what Siri can control. Learn more in Chapter 18, "Controlling Your Smart Home."

Using Siri to Find Interesting and Useful Information

Siri isn't just about controlling apps and settings on your iPad. You can also use Siri to find information—like searching the Web with voice commands.

Find Information

Probably one of the most used functions for Siri is to find information. It's like using Google, but Siri returns a real answer to any question you ask.

You can ask Siri about the weather:

What's the current temperature?
Is it going to rain today?
What's the weather like this weekend?

Or about sports:

What's the score of the Vikings game?
Who's in first place in the National League West?

Or the stock market:

How is the stock market doing today?
What is Apple's stock price?

Interested in a new movie?

Is the new Marvel movie any good?
What time is the Star Trek *movie playing?*

Or just ask Siri about anything in which you're interested. Here are a few queries to try:

How far is it to the moon?
How old is George Clooney?
Who is prime minister of Canada?
Who wrote War and Peace?

Solve Equations and Make Conversions

Siri is also great for converting various measurements. Here's a sampling of what you can ask:

How many kilometers are there in a mile?
How many dollars in a Euro?
Convert 2.7 ounces to grams.
Convert 47 dollars to pounds.

Conversions are fun, but what about basic math problems? Siri can handle all sorts of equations, such as

What is a 20 percent tip on 83 dollars?
What is 238 times 72?
What is 22 divided by 2?
What is the square root of pi?

Find Businesses and Get Directions

Siri can help you find nearby businesses (and information about them). Say something like

Where is the nearest gas station?
Are there any good Chinese restaurants nearby?
How late is Home Depot open?
Is CVS open right now?

You can also use Siri to find and make reservations via the Yelp and OpenTable apps:

Find me a table for two for dinner tomorrow.
Book me a table for four at noon on Thursday at The Fancy Pear.

Then there are directions, of which you can ask Siri:

Show me how to get to the history museum.
Find directions to 10223 Main Street.

You can even ask Siri other travel-related questions, such as

Check flight status of United 231.
When does Southwest 72 arrive?

Triggering Multiple Actions with Shortcuts

To get even more out of Siri, you can create Shortcuts that combine multiple voice commands, notifications, and the things that Siri learns about you into a single command. For example, you can create a Shortcut called "Travel Plans" that, when run, recites or displays hotel and flight information for an upcoming trip. Or you can create custom messages as Shortcuts so you can send texts to recipients with a single command ("send Bob my Welcome text").

Note that Siri Shortcuts can run across multiple apps and multiple Apple devices, including the Apple TV streaming media box and Apple's new HomePod smart speaker. Some Shortcuts are available from Apple and third-party developers; you can also create your own Shortcuts.

Use an Existing a Shortcut

You create and manage Shortcuts with the Shortcuts app.

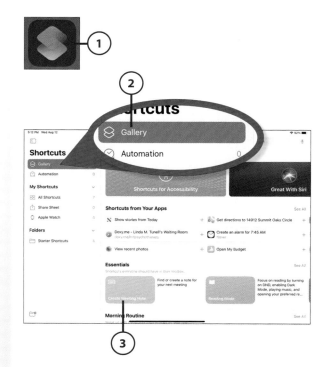

(1) Tap the Shortcuts icon on the Home screen to open the Shortcuts app.

(2) In the Shortcuts sidebar on the left side of the screen, tap Gallery to view preexisting Shortcuts.

(3) Tap any Shortcut you want to use on your iPad.

(4) Tap the right arrow in the Do section to change any of the actions for this Shortcut.

(5) Tap Add Shortcut.

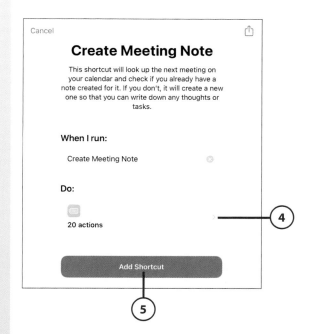

Create a New Shortcut

You can also use the Shortcuts app to create new Shortcuts from scratch.

(1) From within the Shortcuts app, go to the My Shortcuts section of the sidebar and tap All Shortcuts.

(2) All existing Shortcuts are displayed. Tap the three-dot icon (…) on a Shortcut to edit or delete it.

(3) Tap the + icon to create a new Shortcut.

(4) Tap an action in the right pane to add it to the Shortcut.

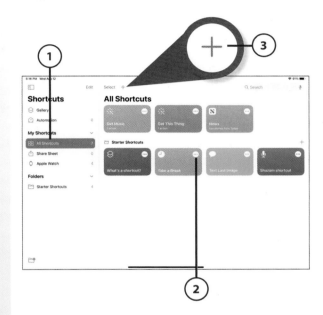

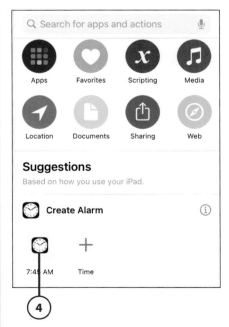

(5) Enter appropriate detail into the action you added.

(6) Tap additional actions as needed to complete this Shortcut.

(7) Tap the Info (…) button next to the name New Shortcut to display the Details panel.

(8) Enter a name for this Shortcut into the Shortcut Name field.

(9) Tap Add to Home Screen to add an icon for this Shortcut to your iPad's Home screen.

(10) Tap Done to create the Shortcut.

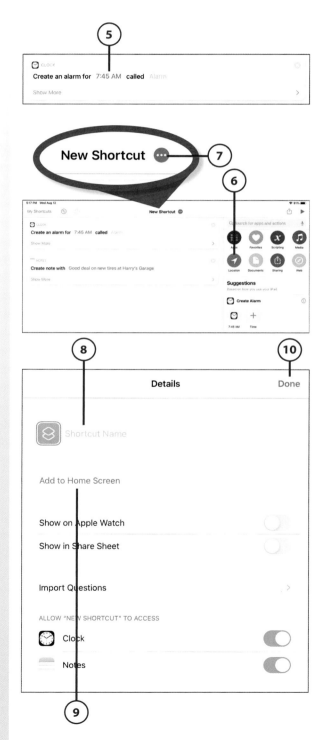

Run a Shortcut

It's easy to run any Shortcut you've added or created. Simply do this:

(1) Launch Siri, either by pressing and holding the Home or On/Off button or speaking "Hey, Siri."

(2) Speak the name of the Shortcut.

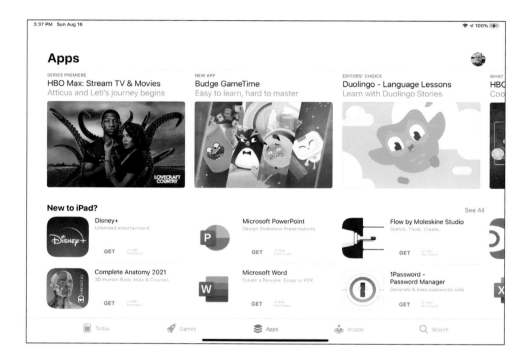

In this chapter, you learn how to find, purchase, install, and use apps on your iPad.

→ Working with Apps
→ Finding New Apps in the App Store
→ Managing Installed Apps
→ Multitasking on Your iPad

Installing and Using Apps

Most of what you do on your iPad you do via *applications*, or *apps*. An app is a self-contained program designed to perform a particular task or serve a specific purpose. There are apps for news and weather, apps for email and text messaging, apps for Facebook and Pinterest, even apps for listening to music and watching videos. Whatever you want to do on your iPad, there's probably an app for it.

Your new iPad came with more than a dozen apps preinstalled, but these aren't the only apps you can use. There are tens of thousands of additional apps available, most for free or low cost, in Apple's online App Store. It's easy to find new apps and install them on your iPad—and then use them every day.

Working with Apps

To do just about anything on your iPad, you have to learn how to work with apps. All the apps currently installed on your iPad are displayed on the various Home screens. Each icon on the screen represents a different app. When you install a new app, an icon for that app appears on the Home screen.

Launch an App from the Home Screen

You open apps from the iPad's Home screen.

(1) Swipe up from the bottom of the screen or press the Home button (if your iPad has one) to return to the main Home screen.

(2) Navigate to the Home screen that displays the icon for the app you want to open, and then tap that icon to open the app.

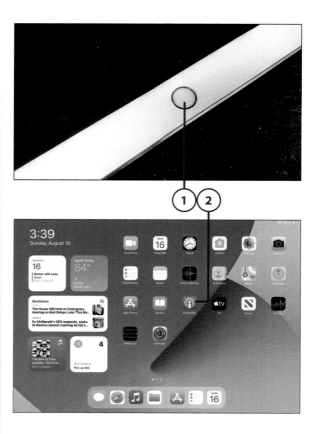

Launch an App from the Dock

You can launch apps from the Dock, which floats at the bottom of every screen on your iPad. The left side of the Dock hosts icons for apps you've placed there. The right side of the Dock displays icons for your most recently used apps. (Chapter 3, "Personalizing the Way Your iPad Looks and Works," explains how to customize the apps on the Dock.)

(1) The Dock floats over the bottom of every Home screen and many apps. If the Dock is hidden, display it by swiping up slightly from the bottom of the screen.

(2) Tap an icon on the Dock to open that app.

3 If you've recently viewed files or documents with a given productivity app, long-press the app's icon on the Dock to view a panel that displays those files.

4 Tap a file to open that file within the app.

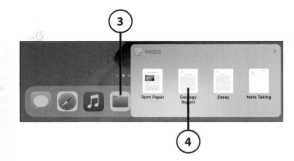

Search for an App

If you have a ton of apps installed on your iPad, it may be challenging to find a specific app you want. (Which Home screen did you put it on?) If this is the case, use your iPad's search function to find that one app you're looking for.

1 Swipe down from anywhere on the Home screen (except the very top) to display the Search pane.

2 Enter the name of the app you want into the Search box.

3 As you type, your iPad suggests matching apps (along with other information). Tap the name of an app to open it.

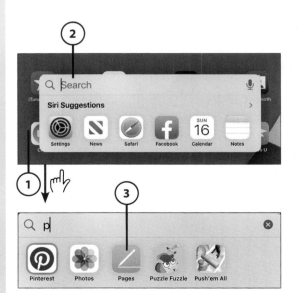

>>>Go Further

UNIVERSAL SEARCH

In iPadOS 14, this onscreen search function has been significantly improved over previous versions. You can now search from within any app—and for more than just apps.

To search from within an app, you need to be using some sort of universal keyboard. Press Cmd+Space to display the Search pane. Otherwise, swipe down from the middle of any Home screen to display the Search pane.

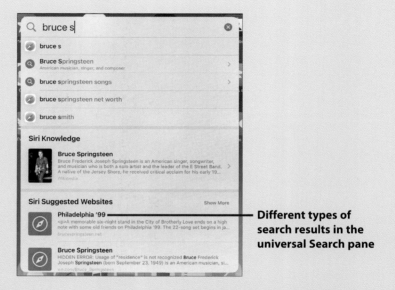

Different types of search results in the universal Search pane

The results you see in the Search pane include apps installed on your iPad as well as settings, files, web pages, news headlines, general information (dubbed "Siri Knowledge"), maps, and music, movies, and TV shows from the iTunes store. Tap a result to launch an app, go to a web page, or see more information.

Switch Between Apps

You can have multiple apps open at the same time and easily switch between them.

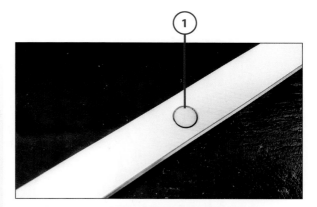

(**1**) On any iPad with a Home button, press the Home button twice. *Or…*

(**2**) Swipe up from the bottom edge of the screen, then swipe to the right, and then swipe up again.

(**3**) You see the App Switcher, which displays all your open apps. Swipe left or right to focus on other apps.

(**4**) Tap an app to switch to that app.

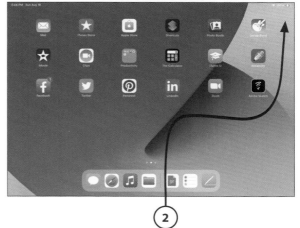

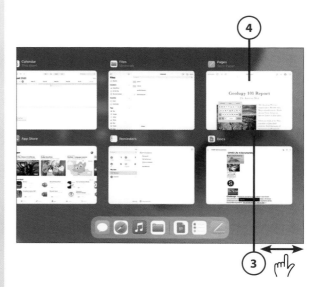

Close an App

Apps remain open until you manually close them. When you're not using an app, it remains paused in the background, but it doesn't slow down your iPad or drain your battery. Because of this, you don't have to close an app when you're done with it—although you can if you want.

1. Press the Home button twice or swipe up from the bottom edge of the screen, then swipe to the right, and then swipe up again.

2. You now see the App Switcher, with all open apps displayed. Tap and drag the app up and off the screen until it disappears.

Finding New Apps in the App Store

Where do you find new apps to use on your iPad? There's one central source that offers apps from multiple developers—Apple's App Store.

Browse and Search the App Store

Apple's App Store is an online store that offers apps and games for iPads and iPhones. Most apps in the App Store are free or relatively low cost. It's easy to find new apps by either browsing or searching.

1. Tap the App Store icon to open the App Store.

2. Tap the Today tab to view stories and recommendations about featured apps and games.

3. Tap the Games tab to view new, top, and featured games for your iPad.

4. Tap the Arcade tab to access the Apple Arcade subscription gaming service.

Apple Arcade

Apple Arcade is a subscription gaming service available to users of Apple iPads, iPhones, and Mac computers. Learn more about Apple Arcade in Chapter 23, "Playing Games."

5 Tap the Search tab to search for apps or games by name.

6 Tap the Apps tab to view new, top, and featured apps.

7 Swipe from right to left to see more apps/games in each list.

8 Scroll down to the Top Categories section to view apps/games by category.

9 Tap See All to view all categories.

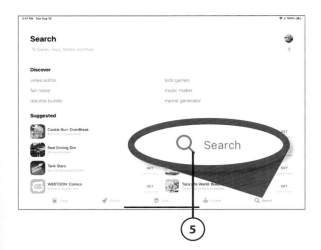

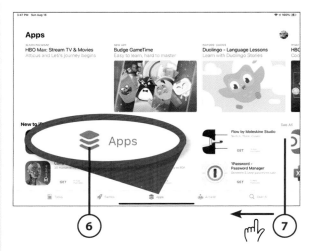

(10) Tap a category to view apps/ games in that category.

(11) Tap an app/game to view that item's app panel.

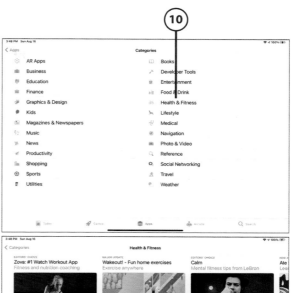

Purchase and Download Apps

Many apps in the App Store are free. Others you have to pay for.

(1) Download a free app from the app panel by tapping the Get button. (If prompted, enter your Apple ID password.)

(2) You see the details panel for that app; tap Install and, if prompted, enter the password for your Apple ID. The app is downloaded to and installed on your iPad.

Locating and Moving Apps

When you install a new app, the icon for that app appears in the first empty space on the second or later Home screen. (New apps are not added to your main Home screen.) You can then move that icon to another position or screen if you like. (Learn more about moving and managing app icons in Chapter 3.)

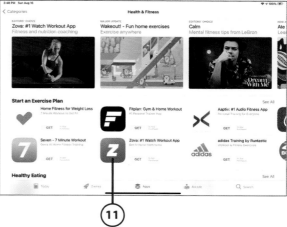

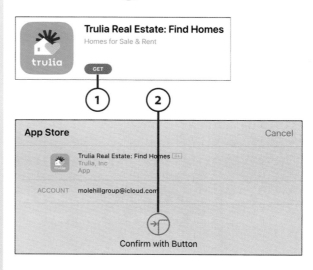

3 Instead of a Get button, a paid app displays a price button. Tap this button to initiate the purchase.

4 The next panel describes how to make the purchase. You may be prompted to confirm within that panel via a Purchase button or by double-pressing the On/Off button. Follow the instructions to make the purchase.

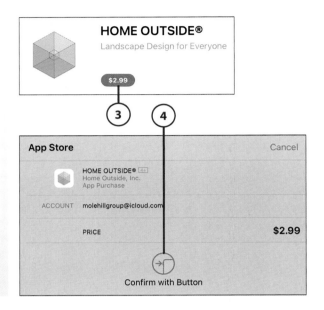

>>>Go Further
PAYING FOR APPS

Naturally, you have to pay for those apps that have a price. You may have provided credit card information when you first set up your iPad or created your Apple ID. If not, you can add a credit card to your account at any time by using the Safari app to go to https://appleid.apple.com and signing in to your Apple account. On the main page, scroll down to the Payment and Shipping section and tap Edit Payment Information. Enter your credit card number, expiration date, billing address, and the like, and then tap Save. This credit card will be used for all your App Store purchases. You don't have to store a credit card to make purchases. You can choose to manually enter your credit card information every time you make a purchase rather than storing the card info in your account.

Managing Installed Apps

Your iPad offers a variety of functions you can employ to better manage the apps you have installed on your device.

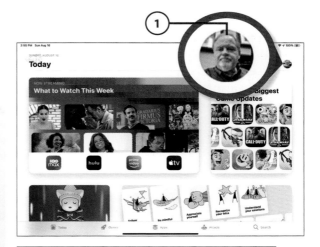

View and Manage Purchased Apps

All the apps you've ever purchased are listed in the App Store app. Even if you've deleted an app from your device, you can still see that app and reinstall it on your iPad at any time.

1 From the Today tab in the App Store, tap the Profile icon in the top-right corner to display the Account panel.

2 Tap Purchased.

3 Tap the All tab to view all the apps you've purchased.

4 Apps that are currently installed on this device have an Open button. Tap the Open button to launch that app.

(5) The App Store keeps track of apps you've installed on other devices (such as your iPhone) that you can also install on your iPad. Tap the Not on This iPad tab.

(6) Apps you've purchased but have yet to install or that you've since deleted (so they're not currently installed) have a cloud icon. Tap this Download icon to install an app on your device.

App Updates

From time to time, the apps on your iPad get updated with new features or bug fixes. By default, apps update automatically, so you'll always have the latest version installed. To check on recent updates, open the App Store app, and tap the profile picture to display the Account pane. Scroll down the pane to view list of Updated Recently apps.

Delete an App

Over time, you'll probably find that you've installed some apps that you no longer use. You may want to delete these apps to clear up any Home screen clutter and to free up storage space for new apps.

(1) Navigate to the Home screen that contains the app you want to delete; then press and hold the app you want to delete.

(2) Tap Delete App.

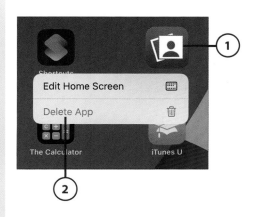

(**3**) Tap Delete to confirm the deletion.

Delete "Photo Booth"?

Deleting this app will also delete its data.

Cancel   Delete —— (**3**)

Can't Delete

Not all apps can be deleted. In particular, you can't delete many of Apple's pre-installed apps. If you don't see a Delete App option for an app, it's there to stay.

Multitasking on Your iPad

You can have multiple apps open simultaneously on your iPad. Normally, you use one app full screen while the other apps hide in the background; you then use the App Switcher to change from one app to another.

You also can choose to run two or more apps (or two or more windows from the same app) onscreen at the same time. Apple's iPadOS includes several features that make it easy to multitask on your iPad. This multitasking works best when you position your iPad horizontally so that it looks like a traditional computer screen. This gives all open apps enough room to operate side-by-side.

Display Multiple Windows with Split View

The Split View feature lets you display and use two apps, or two windows from the same app, at the same time. The apps appear in two equally sized panels on the screen.

(**1**) With the first app open, swipe up slightly from the bottom of the screen to display the Dock.

(2) Tap and drag the icon for the second app from the Dock to the right edge of the screen. Keep holding the app icon until the first app shrinks to the left. You can now drop the icon for the second app into the blank space at the right side of the screen.

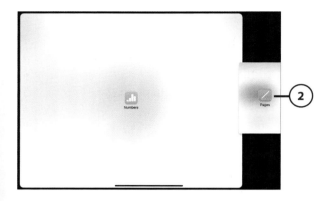

(3) The two apps are displayed side by side. To change the space assigned to each app, drag the handle between the two apps left or right, as necessary.

(4) To turn a Split View app into a Slide Over panel (discussed next), drag the handle for that app down until that app becomes a floating window, then release.

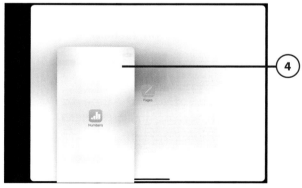

(5) To return to single-app view, drag the app divider handle to the right edge of the screen (to close the right-hand app) or to the left (to close the left-hand app).

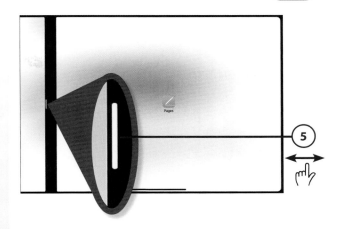

Open a Floating Window with Slide Over

The Slide Over feature enables you to open a second app onscreen without closing the one you're currently in. The second app (called the side app) appears in a separate panel or window that floats on top of the first app.

(1) With the first app open, swipe up slightly from the bottom of the screen to display the Dock.

(2) Drag the icon for the app you want from the Dock up onto the main screen, then release.

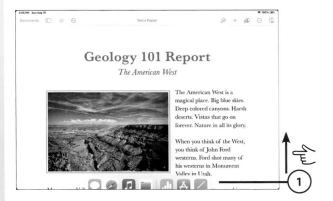

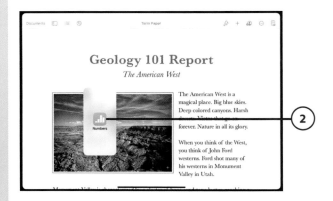

3 You can interact with the main app and the Slide Over app at the same time; they're both "live."

4 Sometimes the Slide Over panel hovers over important information on the full-screen app, and you might need to move it to use the other app. To move the Slide Over panel to the other side of the screen, press the handle at the top of the panel and drag it to the other side.

5 To display a different app in the Slide Over panel, tap and drag that app's icon from the Dock onto the screen. You can have multiple Slide Over windows open at the same time, although only the top one is visible.

6 To view all open Slide Over windows in the Slide Over Switcher, touch and drag the Slide Over window's *bottom* handle upward. Keep your finger on the handle.

7 When you see the other Slide Over windows "peek" through, release your finger.

8. All the open Slide Over windows now appear, spread out in "cards." Scroll through the windows by swiping left and right.

9. Tap the window you want to bring to the front.

10. To close one of these Slide Over windows, drag it up and off the top of the screen.

11. To turn the current Slide Over window into a Split View app, drag it to the left or right edge of the screen and hold it there until the Split View appears.

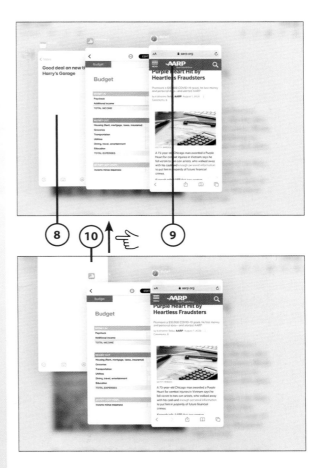

8   10   9

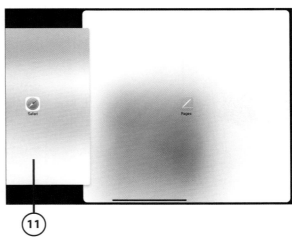

11

12 To display the Slide Over window full screen, drag it to the center top of the screen.

13 To hide the Slide Over panel, press the handle at the top of the panel and quickly drag it off the right side of the screen, without holding it there.

14 To redisplay the Slide Over panel, swipe in from the right side of the screen.

Split View and Slide Over

To display more than two app windows at the same time, open two apps in Split View and then use Slide Over to add a third floating window on top of them.

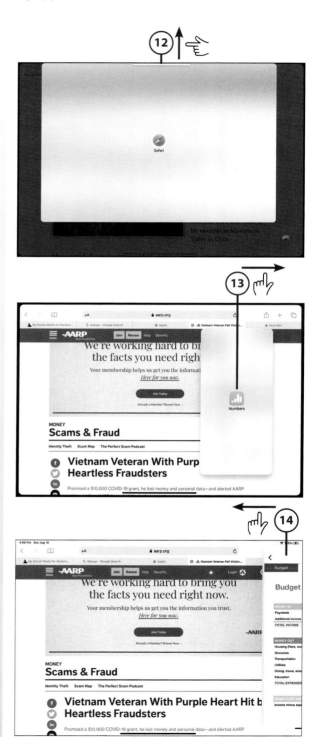

View All Open Windows with App Exposé

With App Exposé, you can quickly view all open windows for any given app. This is particularly useful if you're viewing multiple web pages or multiple word processing or spreadsheet documents. Here's how it works.

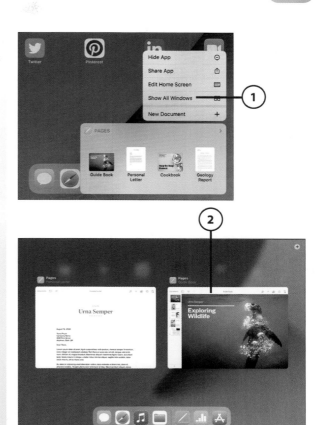

1 Display the Dock and tap and hold the icon for the open app, then tap Show All Windows.

2 All open windows for that app are displayed. Tap a window to display that instance full screen.

Drag and Drop Text and Pictures Between Apps

Your iPad's multitasking features let you easily copy text and pictures between apps using either Slide Over or Split View. (Personally, I find Split View easier for these tasks.) You do this with simple drag and drop.

1 In either Slide Over or Split View, press to select the text or picture you want to copy.

2 Still holding down the selection, drag it to the point in the other app where you want to copy it.

3 Release your finger, and the item is placed in the second app.

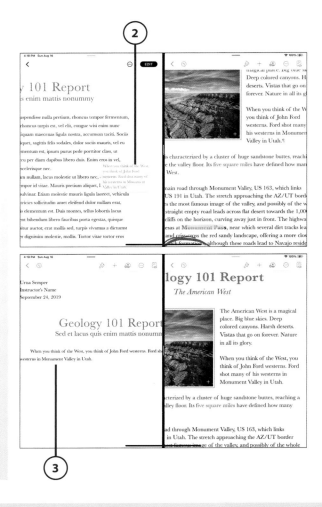

>>>Go Further
PICTURE IN PICTURE

Your iPad also offers a picture-in-picture mode for use with certain apps. If you're watching a TV show or movie, or conducting a FaceTime video chat, you can keep that video or chat playing in a separate window while you do other stuff on the iPad screen. The video or chat appears in a small window that floats on top of the active app, so you can keep watching while you work.

Picture in picture window

To activate picture-in-picture mode, tap the Picture in Picture icon (in the video app), press the iPad's Home button (if your iPad has one), swipe up from the bottom of the video that is playing, or pinch the screen to minimize the app. The video screen shrinks to a corner of your display. Your Home screen or other apps appear beneath the video window, and you can then use whatever app you want while the video or chat keeps playing.

When you display the picture-in-picture window, you can make the window bigger by expanding two fingers on the window. Pinch two fingers together to make the video window smaller. And, if you want, you can use your finger to drag the video window to a different corner or position on the screen.

To return the video window to full screen, tap the window to display the control icons and then tap the Picture in Picture icon.

‹ Mailboxes   Edit

All Inboxes

Q Search   🎤

★ **Ralph Miller**   2:20 PM
Flowers
What do you think of these?

mmiller2007@comcast.n...   2:14 PM
Picture
Here's a great one, don't you think?

★ **Bob Miller**   Yesterday
Zoom meeting invitation - Miles Baker's...
Miles Baker is inviting you to a scheduled
Zoom meeting. Topic: Miles Baker's Zoom M...

Sherry Miller   Yesterday
zoom
Sherry Miller is inviting you to a scheduled
Zoom meeting. Topic: Sherry Miller's Person...

Michael Miller   Monday
Lunch date
Just wanted to see if you were free for lunch
Wednesday next week. I'm thinking Porter Cr...

mmiller2007@comcast....   Monday
New picture
Let me know what you think of this one!

Apple TV   5/25/20

Updated at 2:11 PM

RM ★ **Ralph Miller**   2:20 PM
      To: Michael Miller ›

Flowers

What do you think of these?

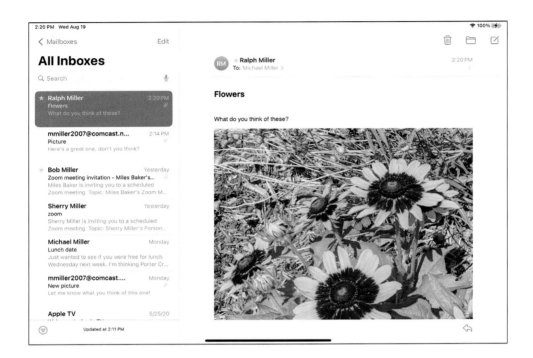

In this chapter, you learn how to use your iPad's Mail app to send and receive email messages.

→ Sending and Receiving Messages
→ Configuring the Mail App

Sending and Receiving Email

For the past several decades, email has been an important means for people to communicate with each other. Whether you're trading messages with friends and family members, receiving messages from groups to which you belong, or receiving confirmation messages after you've purchased merchandise online, it's important to have an email address and master the workings of a full-featured email program.

The email app included with your iPad is called Mail. You can use Mail to send and receive emails using your Apple email account, or you can configure the app to work with other email services, such as Yahoo! Mail and Google's Gmail.

Sending and Receiving Messages

When you set up your iPad with your Apple ID, your account information was automatically added to the Mail app. So if you have an Apple or iCloud email address, that account is already set up in the Mail app—there's nothing more you need to do to get started.

Select an Inbox

If you've configured the Mail app for more than one email account, you have more than one inbox to deal with. The Mail app includes inboxes for each account you use as well as a VIP inbox, which lists messages from people you've decided are very important.

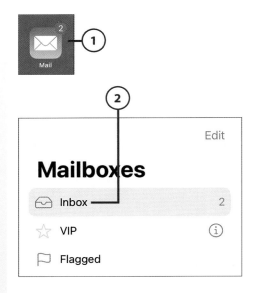

1. Tap the Mail icon to open the Mail app. (If you have unread email, you see a number on the Mail icon, which signifies how many messages you have waiting.)

2. If you only have a single iCloud mail account, you see a single Inbox in the Mailboxes pane. Tap Inbox to view all messages in the inbox. *Or…*

3. If you have multiple email accounts, you see multiple accounts listed. Tap the name of an account to view messages from that specific account. *Or…*

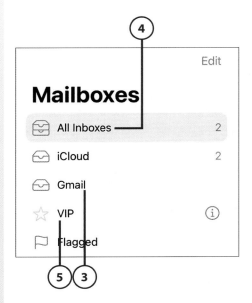

4. Tap All Inboxes to see all your email from all your accounts together. *Or…*

5. Tap VIP to view messages from your most important contacts.

>>>Go Further
ADDING A VIP

To add people to the Mail app's VIP list so that their incoming messages automatically appear in the VIP inbox, tap VIP to open the VIP inbox, and then tap Add VIP. You see a list of all your contacts; tap the name of the person you want to add to your VIP list. After you've added one person, you can add other people by tapping the Information (i) icon next to VIP in the Mailboxes pane and then tapping Add VIP.

Read a Message

Reading a message in the Mail app is subtly different depending on how you hold your iPad. When you hold the device horizontally, you see a split screen with inbox contents on the left and the selected message on the right. When you hold the device vertically, so the screen isn't quite as wide, you see only the selected message; the inbox list slides offscreen to give more room to the selected message.

In this and the following tasks, I show you how Mail works in horizontal mode—which works best for reading emails.

1. From within the Mail app, tap the inbox you want to view. The left column slides offscreen to reveal all the messages in this inbox.

2 Unread messages are displayed with a blue dot next to the subject line; messages you've read don't have this blue dot. Each message displays the sender's name and/or email address, the time or date received, message subject, and the first line or so of the message. Tap the message you want to read; the selected message header is shaded.

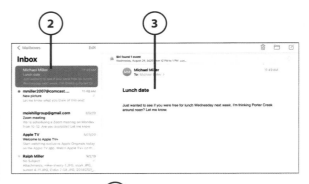

3 The selected message displays on the right side of the screen. (Or, if the iPad is held vertically, it appears full screen.) The name of the sender appears at the top, with the message subject beneath that. The text of the message fills the bottom of the message pane; if this message is in reply to a previous message, the previous message is "quoted" beneath that.

4 If the sender has attached photos to the message, those photos are automatically downloaded and displayed within the message text. (If a photo is not automatically displayed, tap the Download link to display it.)

It's Not All Good

Unwanted Attachments

If you receive an attachment via email that you weren't expecting—especially from someone you don't know—do *not* tap to open it! Attachments can contain viruses and other malware that can infect or damage your iPad. You should delete unwanted file attachments and the emails to which they come attached!

Reply to a Message

When you reply to an email message, the original message is "quoted" beneath your reply.

1. Open the original message and then tap the Reply icon. This displays a panel of options.

2. Tap Reply to open the Reply pane. (Alternatively, tap Reply All to reply to everyone copied on the original message, or Forward to forward this message to someone new.)

3. The original sender is listed as the recipient in the To field, and the subject is also automatically filled in (although you can edit the subject if you want). Enter your reply into the main text box.

4. Tap Send to send your reply to the original sender.

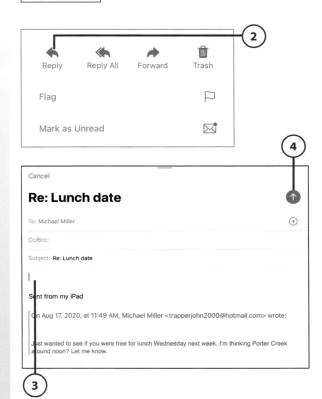

Create and Send a New Message

Writing and sending a new email message is similar to replying to an existing message. All you need to know is who you're sending it to!

1. From within the Mail app, tap the New Message icon at the top-right corner of the screen. This displays the New Message pane.

2. Tap within the To field and start typing the name or email address of the intended recipient.

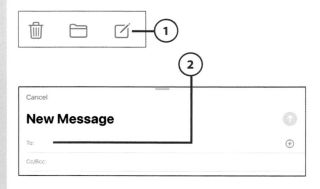

3 As you type, the Mail app displays matching names from your contacts list. If this person is listed there, tap his or her name. If you're sending an email to someone not in your contacts list, continue entering the person's email address manually.

Contacts

Learn more about your contacts and the Contacts app in Chapter 10, "Managing Your Contacts."

4 Add another recipient by tapping again within the To field and repeating steps 2 and 3.

5 Add Cc or Bcc recipients by tapping Cc/Bcc and adding the desired names.

Cc and Bcc

Cc stands for *carbon copy* and sends a copy of your message to additional recipients. Bcc stands for *blind carbon copy* and sends a copy of your message to additional recipients but hides their names and addresses from the main recipients.

6 Tap within the Subject field and enter the subject of this email.

7 Tap within the main text box and enter your email message.

8 Tap Send to send the message.

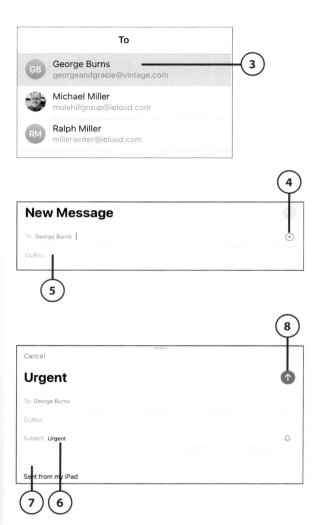

Attach a Photo to a Message

The Mail app lets you attach photos and other files to your email messages. This makes it easy to share pictures with your friends and family.

Attachments

A file of any sort—such as a text document or photo—attached to an email message is called an *attachment*.

① Create a new message as usual, and tap within the main text box to display the icons above the onscreen keyboard; then tap the Photos icon to display the Photos pane.

② Tap All Photos.

③ Tap Photos to display individual photos or Albums to display photo albums.

④ Tap to select the photo you want to attach.

⑤ Tap Use.

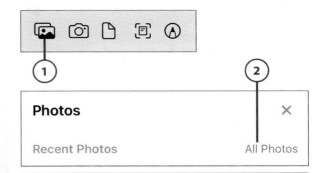

6 The photo is added to your new message. Complete your email as normal and tap Send.

It's Not All Good

Keep It Small

Many digital photos, especially ones you take with your iPad camera, can create fairly large image files. If an image file is too large, your email service (or your recipient's email service) might reject it. For this reason, the Mail app might prompt you to resize any large photo you attach. If you're so prompted, choose a smaller size before proceeding. (Note, however, that a smaller image size won't look as good if printed.)

Delete a Message

When you're done reading a message, you might want to delete it. You might even want to delete a message before you read it if it's from some entity you don't know or don't want to hear from. Fortunately, you can delete messages from the inbox itself or from the reading pane.

1 To delete a message after you've read it, display the message and then tap the Trash icon.

(**2**) To delete one or more messages
from the inbox, tap Edit.

(**3**) Tap to select the message(s) you
want to delete.

(**4**) Tap Trash.

Organize Messages in Files

What do you do with your old messages after you've read them? You might want to delete some of them (and you just learned how), but you might want to keep some around for future reference. Instead of letting old messages clog up your inbox, create folders to organize and store your old messages.

(**1**) From the Mailboxes list, tap Edit.

(**2**) Tap New Mailbox. (The Mail app refers to folders as *mailboxes*.)

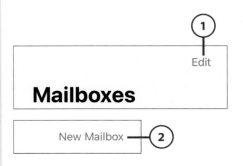

3 Tap within the Name field and enter a name for this new folder/mailbox.

4 Tap Mailbox Location.

5 Tap the account where you want to store this folder.

6 You can also select a master folder for this new folder; the new folder then becomes a subfolder within that folder.

7 Tap Save.

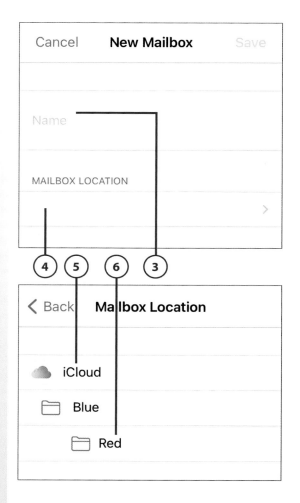

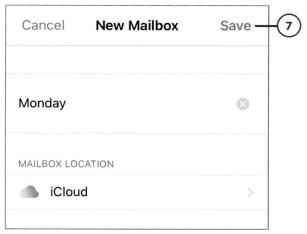

8 Tap Done. The new folder/mailbox is now created.

9 Move an email message to a folder by opening the message and then tapping the Folder icon at the top of the screen.

10 Tap the folder to which you want to move this message. The message is moved to that folder.

11 Tap a folder/mailbox in the Mailboxes pane to view the contents of that folder/mailbox.

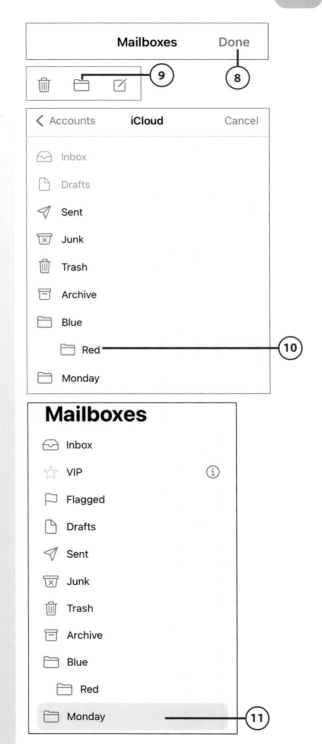

Deal with Junk Email (Spam)

One of the more annoying things about email is the amount of junk email, or spam, you sometimes receive. Fortunately, the Mail app includes a spam filter that attempts to identify junk email and automatically send it to a special Junk E-mail folder.

(1) View messages in the Junk folder by going to a given account in the Mailboxes pane and tapping Junk. (If you find something there that *isn't* spam, you can move it back to your normal inbox.)

(2) Send a spam message you've received to the Junk E-mail folder by opening the message and tapping the Folder icon.

(3) Tap to select the Junk folder.

OTHER EMAIL APPS

Although Apple's Mail app is a very good email app, and it's included free with your iPad, there are other email apps available for your use. If you use a different email service, such as Gmail or Outlook.com, you may want to use the specific app for that service instead. You can find apps for most of the major email services in Apple's App Store; just search by the name of your email service.

Configuring the Mail App

The Email app is automatically configured to work with your Apple iCloud email. It can also work with Microsoft Outlook, Google's Gmail, Yahoo! Mail, and other email services.

Add an Email Account

If you have an email address with a service different from iCloud, you can add that account to the Mail app and check that email from the app. The Mail app lets you add multiple accounts and check messages from multiple addresses and inboxes. It's convenient to be able to check all your emails in one place.

Although you can still go to the Gmail website for your email, for example, it's just as easy to check your Gmail messages from within the Mail app. All you have to do is configure the Mail app for the new account.

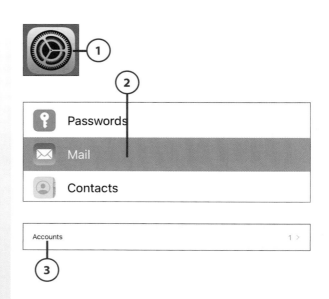

1. Tap the Settings icon to open the Settings screen.

2. Scroll down the left column and tap Mail.

3. Tap Accounts.

4 In the Accounts section, tap Add Account to display the Add Account screen.

5 Tap the type of email account you want to add—iCloud, Microsoft Exchange, Google, Yahoo!, AOL, or Outlook.com. (If you have another type of email, tap Other.)

6 When prompted, sign in to the email account you selected with that account's email address or phone number and password. (Different email providers require different information.)

7 If prompted to enable various aspects of your email account (such as calendar or contacts), select those options you want.

8 Tap Save. This account is now added to the Mail app.

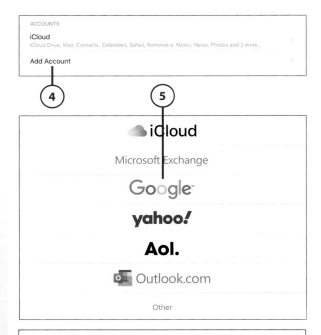

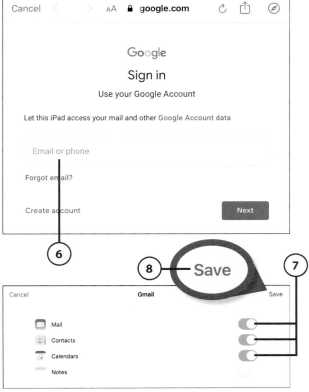

>>>*Go Further*
SPECIFY A DEFAULT ACCOUNT

Although Mail receives messages from all accounts you add, you need to select which account is used when you send messages. This is important if you have, say, both iCloud and Gmail accounts added in the app; you have to decide whether the new emails you send come from your iCloud or Gmail address.

To do this, open the Settings screen, tap Mail, and then scroll down to the Composing section. Tap Default Account and select the account you want to use to send new messages.

You're not stuck with this one account, however. You can select a different sending account when creating a new message. When you create a new message, tap within the From field to display a panel with all your accounts listed, and then select a different email account.

Get Notified of New Messages

The Mail app can notify you when you receive new messages.

1. Tap the Settings icon to open the Settings screen.

2. Tap Notifications in the left column.

3. Scroll down and tap Mail.

4. Tap "on" the Allow Notifications switch.

5. Tap the account from which you want to receive notifications. (You can repeat these steps to config- ure different—or similar—notifi- cations for multiple accounts.)

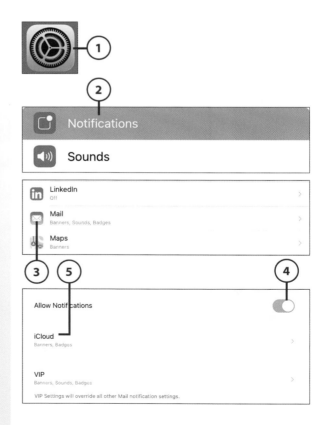

6. Indicate that you want to show notifications from this account on the Lock screen, Notification Center, or Banners by tapping "on" the appropriate option(s).

7. To display previews of message contents in the alerts, tap Show Previews.

8. Select when you want to show message previews: Always, When Unlocked, or Never.

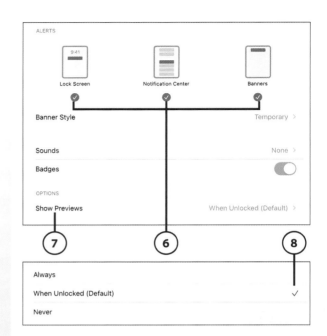

Create a Signature

By default, when you send email from the Mail app, it automatically attaches a line of text to the bottom of each message: "Sent from my iPad." This line is called a *signature*, and you can customize the text.

1. From the Settings screen, scroll down the left column to the list of apps and tap Mail.

2. Scroll down the Mail screen to the Composing section and tap Signature.

3. Tap All Accounts to create a signature for use with all your email accounts. *Or…*

4. Tap Per Account to create a signature for a specific email account.

5. Tap within the text box, delete the default signature ("Sent from my iPad"), and type a new one.

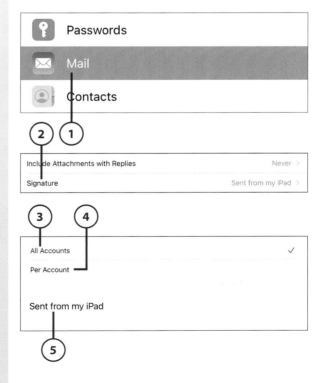

>>>Go Further

SIGNATURES

You should use your email signature to tell your recipients something about you. It can contain useful or fun information.

Many people use their signatures to display additional contact information, such as a phone number or Twitter handle. Others display their job title or professional qualifications in their signatures.

A signature doesn't have to be serious, however. Some people end their emails with a favorite quotation or even a personal statement of some sort. Just keep it short—nobody wants to read a six-line signature!

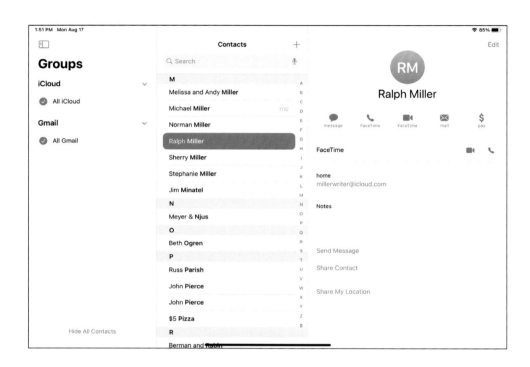

In this chapter, you learn how to manage contacts on your iPad for email, messaging, and video chatting.

→ Using the Contacts App
→ Viewing and Contacting Your Contacts

10

Managing Your Contacts

When you use the Mail app to send an email to someone, you can enter that person's email address manually. That's not a big deal—unless you send a lot of emails to that person, of course, in which case re-entering that same email address each time gets pretty annoying.

It's also annoying to try to remember the email address of each and every person with whom you communicate. Or their phone numbers. Or their street addresses. If you have a lot of friends and family, all that information becomes overwhelming.

Used to be you might keep people's names and addresses in a physical card file or address book. Even better (and more convenient), you can store all your contact information in your iPad's Contacts app. It's a lot easier to have the Contacts app remember all these names and addresses than it is to do it yourself!

Using the Contacts App

The Contacts app is preinstalled on your new iPad. You use the Contacts app to store the names, email addresses, street addresses, phone numbers, and other important information of the people you contact.

Add a New Contact

To add a new contact to the Contacts app, all you need to know is that person's name and any contact information about that person.

1 Tap the Contacts icon to open the Contacts app. All your existing contacts are listed in alphabetical order.

2 Tap the + icon to display the New Contact panel.

3 Enter the person's first and last names into the First Name and Last Name fields. (Or if you're entering a company or organization, enter that entity's name into the Company field.) Past this point, any information you enter is optional—that is, you only need to enter the contact's name to create a new entry.

4 Tap the + next to Add Phone and enter the contact's phone number.

5 Tap the + next to Add Email and enter the contact's email address.

6 Scroll down, tap the + next to Add Address, and enter the contact's address.

7 Add other information by tapping the + next to that field and then entering the appropriate data.

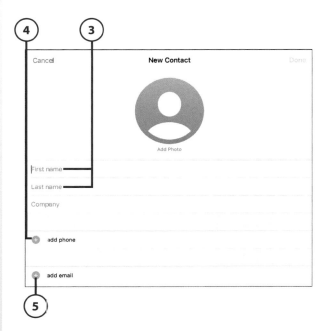

⑧ Include a photo of this person (if you have one) by tapping Add Photo, tapping Choose Photo, and then selecting a picture from your iPad photo library. (Alternatively, tap Add Photo and then tap Take Photo to take a new photo of this person.)

⑨ Tap Done to save the contact.

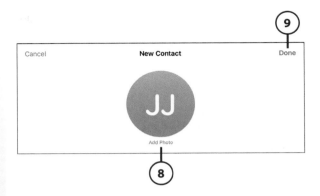

Edit a Contact

Sometimes you might want to add new information about a given contact or change existing info.

① From within the Contacts app, tap the name of the contact you want to edit.

② Tap Edit.

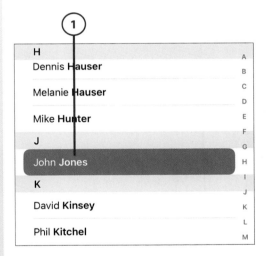

③ Tap within any field you want to edit and make the necessary changes.

④ Tap Done.

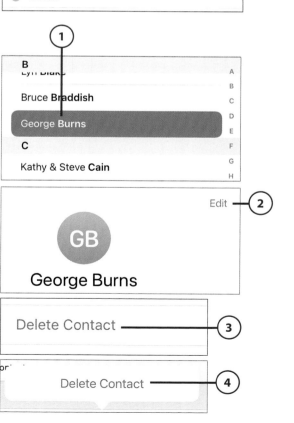

Delete a Contact

You can delete contacts you're no longer interested in.

① From within the Contacts app, tap the name of the contact you want to delete.

② Tap Edit.

③ Scroll to the bottom of the contact information and tap Delete Contact.

④ When asked to confirm this deletion, tap Delete Contact.

Viewing and Contacting Your Contacts

The Contacts app displays the names of all your contacts in the left column. When you select a contact, information for that person or entity is displayed on the right side of the screen.

Display a Contact

By default, your contacts are listed in alphabetical order by last name.

(1) From within the Contacts app, in the contacts list, tap a letter on the right edge of the list to display all contact names that start with that letter.

(2) Tap a contact name to display that contact's information on the right side of the screen.

Search for a Contact

If you have a lot of names in your contacts list, it might be quicker and easier to search for a specific person instead of scrolling through all your contacts.

(1) From within the Contacts app, tap within the Search box. Use the onscreen keyboard to begin typing the person's first or last name.

(2) As you type, the Contacts app displays matching names. Tap to select the contact you want.

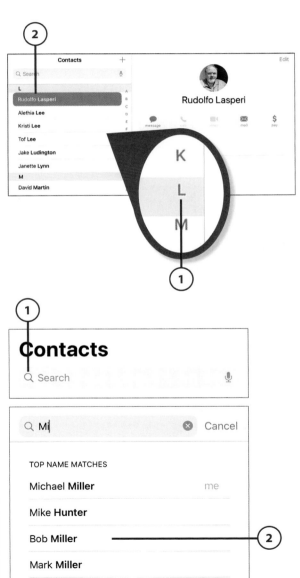

Contact a Contact

Many apps—including Mail, Messages, and FaceTime—use your contacts list to help you easily email or message your contacts. You can also send email to, message, or call (via FaceTime) a person directly from the Contacts app.

1. From within the Contacts app, tap to select the person you want to contact.

2. Tap the Mail icon to send an email to this person. (The icon might say "Home" or "Work," if you've specified that person's home or work email address.)

3. Tap Message to send a text message to this person.

4. Tap the Call icon (the phone) to initiate a FaceTime voice call with this person. (Depending on the information for this contact, the label might say Call, FaceTime, Mobile, or Home.)

5. Tap the Video icon (the video camera) to initiate a FaceTime video call with this person. (This icon is grayed out if the person isn't currently online and capable of receiving FaceTime calls.)

6. Tap the Pay icon to send money to this person via Apple Pay.

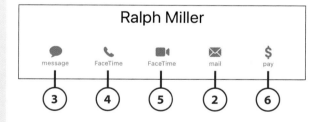

FaceTime and Messaging

To learn more about FaceTime and text messages, turn to Chapter 11, "Video Chatting and Texting."

In this chapter, you learn how to communicate
with friends and family using the Messages
and FaceTime apps.

→ Text Messaging with the Messages App
→ Video Chatting and More with FaceTime
→ Group Video Meetings with Zoom

Video Chatting and Texting

You've learned how to communicate with friends, family, and businesses with email. But email is slow (relatively) and not at all like having a conversation in real time.

For that, we turn to text messaging and video chatting. Text messaging, using the iPad's Messages app, is great for sending short thoughts and notes. Video chatting, using the FaceTime app, is better when you want one-on-one, face-to-face conversation.

Text Messaging with the Messages App

Let's start with text messaging, like you do on your phone. On your iPad, you use the Messages app to send text messages over Wi-Fi via Apple's iMessage service. Your messages can include photos, videos, and other files you attach to the basic text.

Send and Receive Text Messages

The Messages app lets you send text messages to any of your friends and family who have an iPad, iPhone, or Mac computer. It's as easy as tapping out a message on the onscreen keyboard and then tapping the Send button.

1. Tap the Messages icon to open the Messages app.

2. Previous conversations are listed in the left column. Tap a conversation to send another message to this person.

3. Tap the Compose icon to start a new conversation.

4 Tap in the To field and use the onscreen keyboard to begin entering the name, email address, or phone number of the person with whom you want to talk.

5 As you type, the Messages app displays a list of matching people from your contacts list. Tap the name of the person you want to text with or finish entering the contact information for this person.

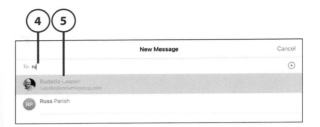

Group Messages

To send a message to a group of people, enter multiple names into the To box. Everyone listed here will receive and be able to respond to your message—and all subsequent responses.

6 Type your message into the iMessage box.

7 Tap the blue Send icon.

Emoji

To include an emoji character in your text, tap the Emoji (smiley face) button on the onscreen keyboard and make a selection from there.

8 Your message appears on the right side of the message pane.

9 Messages from the other person appear on the left. Continue typing to continue the conversation.

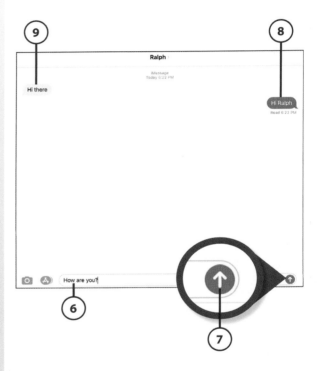

Send an Audio Message

Instead of sending a plain text message, you can record and send a short audio message. Just tap the Record button and start talking!

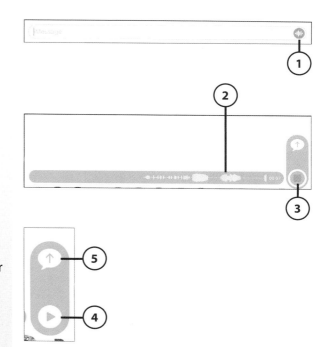

1. In the iMessage box, press and hold the record icon.

2. Begin talking. While you talk, continue to hold your finger on the icon. You see the waveform of your voice in the iMessage box.

3. When you're done talking, lift your finger from the screen.

4. Tap the Play button to hear what you recorded.

5. Tap the Send button to send the recording to the other person.

Send a Photo

Just as with phone-based MMS (multimedia messaging), you can send photos to people you're messaging in the Messages app. You can send photos stored in your iPad's photo library or take a new photo (like a selfie) and send it.

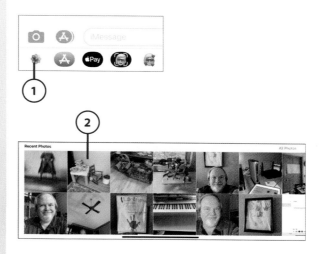

1. To send a photo stored on your iPad, tap the Photos icon below the iMessage box. (You might need to tap the Apps icon to display icons for Photos and other apps.)

2. Navigate to and tap the photo you want to send. (You may need to tap Add to add the photo.)

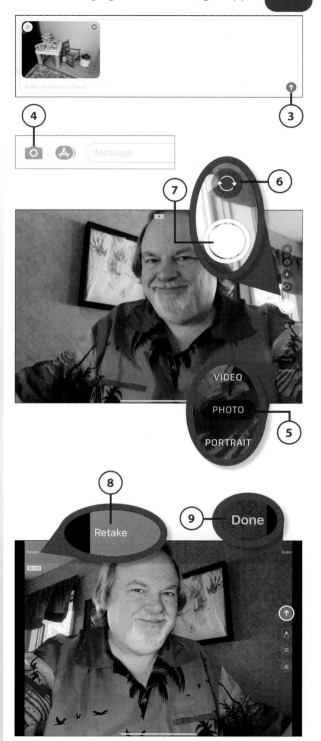

3 The photo is added to a message. Tap the Send icon to send the photo(s) to the person with whom you're texting.

4 To take and send a new photo, tap the Camera icon. This displays the camera screen.

5 Make sure that Photo is selected.

6 If necessary, tap the Flip icon to switch from the rear camera to the front camera.

7 Aim your iPad and then tap the big round icon to take the picture.

8 You see the picture you just took. If you don't like it, tap Retake.

9 If you do like the photo, tap the Done icon.

10 The picture is added to the message. Tap the Send icon to send the photo.

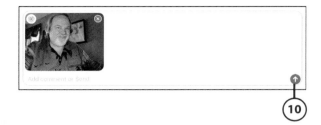

Send a Video Message

Just as you can send photos and short audio messages, you can also record and send short video messages, using your iPad's front-facing FaceTime camera.

1 From within the Messages app, tap the Camera icon next to the iMessage box to display the camera screen.

2 Tap Video.

3 If necessary, tap the Flip icon to switch to the front-facing camera.

4 Look into the camera and tap the red Record button. Start talking!

5 When you're done talking, tap the Record button again to stop the recording.

6. You now see the video you recorded. Tap the Play button to play the video if you want.

7. If you like the video you recorded, tap the Send icon to send the message to the recipient.

Video Chatting and More with FaceTime

Text messaging is nice, but sometimes you want a face-to-face conversation with someone. Maybe you're snowbirding down south while your grandkids are still in school up north. Maybe your siblings or friends live in a different part of the country than you do. Maybe you or one of your family members is stuck at home and you just want to check in or keep in touch.

When a text message or phone call isn't good enough, use Apple's FaceTime app to video chat in real time with your friends and family. All you need is your iPad—and its built-in camera and microphone. Every call you make is free!

Apple Users Only

FaceTime is only available on Apple devices—iPads, iPhones, and Mac computers. To chat with someone who doesn't have an Apple device, you have to use another video chat app, such as Skype or Zoom.

Start a Video Chat

You can use FaceTime to video chat with anyone who has an iPad, iPhone, or Mac that includes a front-facing camera.

1. Tap the FaceTime icon to open the FaceTime app.

2. You see yourself, via your iPad's front-facing FaceTime camera. Overlaid on the left side of the screen is a list of people with whom you've recently chatted. To chat again with one of the people listed, tap the Information (i) icon next to his or her name.

3. You now see the contact panel for that person. Tap the FaceTime video icon. *Or…*

4. To chat with someone new, tap the + icon.

5 Start typing a person's name into the To: box.

6 As you type, matching contacts are displayed. Select the person you want from this list, or finish typing the person's email address or phone number.

7 Tap the Video button, and FaceTime dials the other person.

8 If that person is available and answers, you see that person onscreen. A small thumbnail in the corner of the screen displays what the other person is seeing (you). Have a nice conversation!

9 Tap Mute to mute the sound. Tap the button again to resume normal conversation.

10 Tap the red End icon to end the chat.

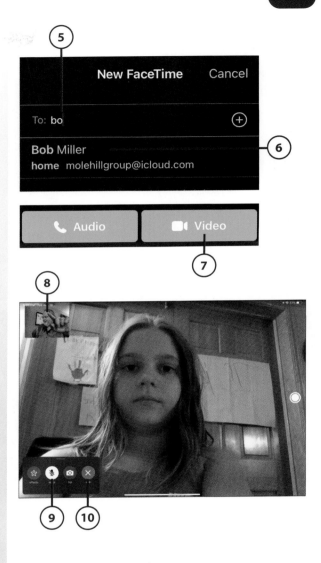

Add a Person to a Group Video Chat

FaceTime also lets you create group video chats, so you can chat with several people at once.

1 From within a FaceTime video chat, drag the top of the control panel up to expand it.

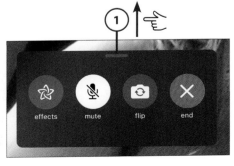

(2) Tap Add Person.

(3) Enter the name or email address of the person you want to add.

(4) Tap Add Person to FaceTime.

(5) FaceTime splits the screen and dials the other person.

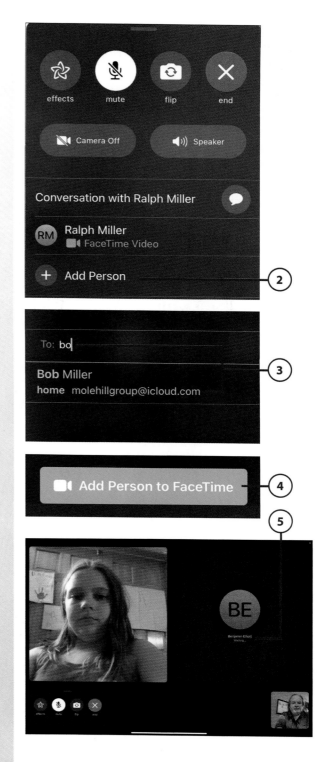

Answer a FaceTime Request

Just as you can call others via FaceTime, they can call you. How do you answer an audio or video call? By pressing an icon and talking, of course!

1. When someone calls you via FaceTime, you see a FaceTime panel at the top of the screen. Tap Accept to accept the call. *Or…*

2. Tap Decline to reject the call.

Make an Audio Call

FaceTime also lets you participate in audio calls—like phone calls but from your iPad to other Apple devices. This is good if you just want to talk and don't need to see the other person, if you're not quite presentable enough to show yourself to the other person, or if you're talking to someone who doesn't have a front-facing camera in his or her device.

1. From within FaceTime, either tap a person's name or tap + to start a new conversation with someone.

2. Tap the FaceTime Audio button. (If this contact has both a FaceTime account and phone number listed, you may see an option for Call or FaceTime Audio. Only select Call if you want to call the mobile or landline number.)

3 FaceTime dials the other person. If that person is available and answers, you see the FaceTime controls and a timer for the call. Start talking!

4 Tap the Mute icon to mute the sound. Tap the icon again to resume normal conversation.

5 Tap the FaceTime video icon to turn this audio call into a video call.

6 Tap the red End icon to end this call.

Group Video Meetings with Zoom

During the coronavirus shutdown, many people kept in touch with co-workers, family members, book clubs, and school classes with Zoom, a popular service for group video meetings. Zoom meetings can be joined from any device with an Internet connection and a web browser; iPads have proven particularly popular with users for their Zoom meetings.

The Zoom app is free from the App Store.

Cross Platform

Unlike FaceTime, Zoom works across all computer and mobile platforms. This way people with virtually any device can participate in a Zoom meeting—iPads, iPhones, Mac and Windows computers, Android phones and tablets, and even Linux and Chromebook computers.

Join a Meeting from a Link

If someone is hosting a Zoom meeting, they'll send you an invitation to join that meeting. You can join the meeting by clicking the link in the invitation or by entering meeting information into the Zoom app (discussed next).

1. Tap the Zoom icon to launch the Zoom app.

2. In the email message, tap the link to the meeting. Upon entering the passcode from the email, the host will now admit you to the meeting.

Join a Meeting from Within the Zoom App

Sometimes a Zoom invitation includes a meeting ID and password rather than a link to the meeting. You can manually enter these into the Zoom app to join a meeting.

1. Tap the Zoom icon to launch the Zoom app.

2. From within the Zoom app, tap Join.

3. Enter the meeting ID into the Meeting ID field.

4. Tap Join.

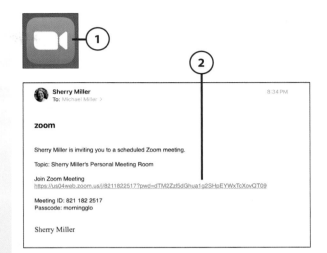

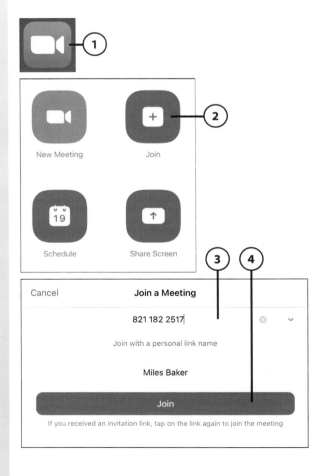

5 When prompted, enter the password.

6 Tap Continue. The host will admit you to the meeting.

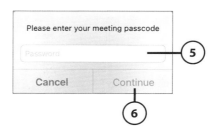

Participate in a Zoom Meeting

It's easy to participate in a Zoom meeting. You'll see the other participants on screen, and they'll see you, assuming video is turned on. All you have to do is talk.

1 By default, Zoom displays in Active Speaker view. The current speaker occupies most of the screen, and you appear in a smaller window. You do not see the other participants unless they're speaking.

2 Tap the screen and then tap Switch to Gallery View to switch to Gallery View.

3 In Gallery View, you see thumbnails for every participant, arranged in a grid pattern. The current speaker's thumbnail appears with a green border.

4 Tap the screen and then tap Mute to mute your sound. Tap Unmute (the same icon) to unmute your sound.

5 Tap the screen and then tap Stop Video to turn off your camera. Tap Start Video (same icon) to turn your camera back on.

6 Tap Participants to see a list of participants in this meeting.

7 Tap Switch to Active Speaker to switch back to Active Speaker view.

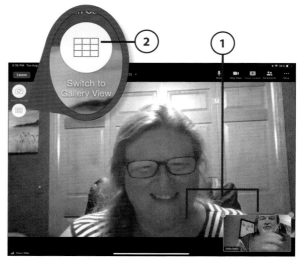

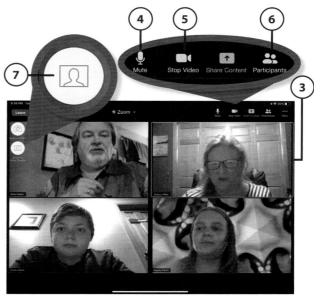

8. To privately text chat with one or more participants, tap More (…) and then tap Chat. This displays a Chat pane.

9. Pull down the Send To list and choose who you want to chat with. Select Everyone to send the message to all participants.

10. Type your message in the message box and then tap Send.

11. To change the background that appears behind you, tap More (…) and then tap Virtual Background.

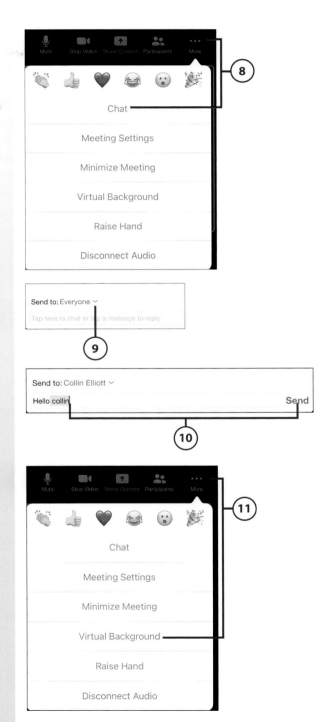

12 Tap the desired background. You see how you look against that background. *Or…*

13 Tap + to select one of your own photos as a background.

14 Tap None to not use a background.

15 Tap the X to return to the meeting.

16 When the meeting is over, or when you need to leave, tap Leave.

17 When prompted, tap Leave Meeting.

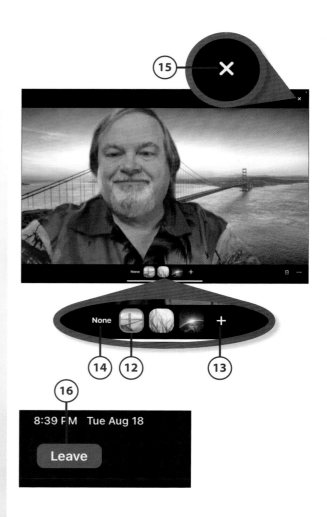

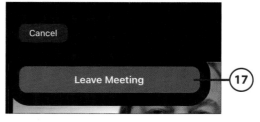

Start a New Zoom Meeting

With a Zoom account, it's easy to start your own meetings. You can start a new meeting at any time or schedule a meeting in advance (discussed next).

Free Limits

In the free version of Zoom, your meetings can last up to 40 minutes and include up to 100 participants. If you want to hold longer meetings, subscribe to the Pro level at $14.99 per month. Learn more at www.zoom.us.

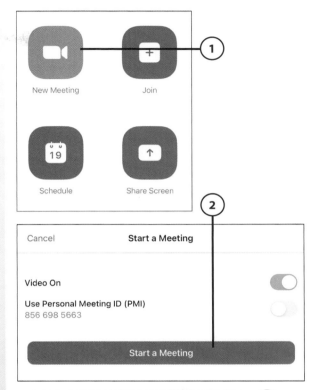

(1) From within the Zoom app, tap New Meeting to immediately start a meeting.

(2) In the Start a Meeting panel, make sure your video is turned on and then tap Start a Meeting.

(3) The meeting is now live. Tap Participants to invite others to the meeting.

(4) Tap Invite.

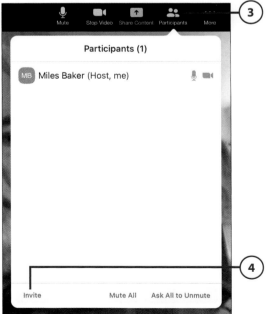

5) Tap Send Email to email invitations. Alternatively, you can invite others via text message (Send Message), send invitations direct to people on your contact list (Invite Contacts), or copy the invitation link so you can paste it into your own email or message (Copy Invite Link).

6) Enter one or more email addresses into the To field.

7) Tap in the message field and enter your own message, if you like, in addition to the automatically generated link, meeting ID, and passcode.

8) Tap Send.

9) When a participant responds, you see a notice that they've entered the "waiting room." Tap Admit to let them into the meeting.

10 Conduct the meeting as usual.
Tap the screen and then tap End
to end the meeting.

11 When prompted, tap End
Meeting for All. (If you're just leav-
ing the meeting and others are
continuing, tap Leave Meeting
instead.)

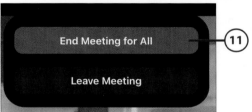

Schedule a Meeting in Advance

In most instances, you'll want to
schedule meetings in advance, to give
participants a chance to prepare.

1 From within the Zoom app, tap
Schedule.

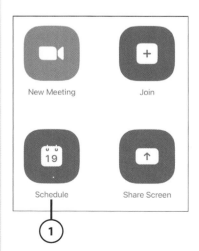

(2) In the Schedule Meeting panel, tap within the top field and give the meeting a name.

(3) Tap Starts and enter a start time and date.

(4) Tap Duration and enter how long the meeting will be.

(5) Tap Save.

(6) When the New Event panel appears, tap Add.

(7) From the main screen, tap Meetings in the left sidebar.

(8) This displays all scheduled meetings. Tap the meeting you just added.

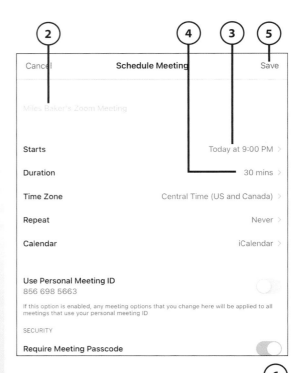

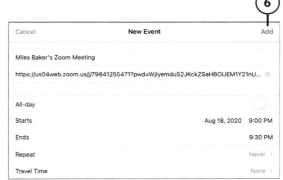

9 Tap Add Invitees.

10 Tap Send Email. (Or, to send an invitation via text message, tap Send Message.)

11 Enter one or more email addresses into the To field.

12 Tap in the message field and enter your own message, if you like, in addition to the automatically generated link, meeting ID, and passcode.

13 Tap Send.

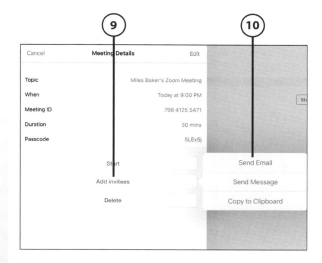

9

10

Cancel   Meeting Details   Edit

Topic   Miles Baker's Zoom Meeting
When   Today at 9:00 PM
Meeting ID   798 4125 5471
Duration   30 mins
Passcode   5LEx5j

Start   Send Email

Add invitees   Send Message

Delete   Copy to Clipboard

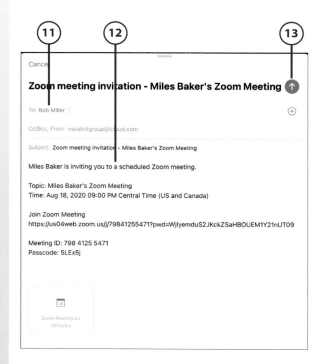

11   **12**   **13**

Cancel

Zoom meeting invitation - Miles Baker's Zoom Meeting

To: Bob Miller

Cc/Bcc, From: molehillgroup@icloud.com

Subject: Zoom meeting invitation - Miles Baker's Zoom Meeting

Miles Baker is inviting you to a scheduled Zoom meeting.

Topic: Miles Baker's Zoom Meeting
Time: Aug 18, 2020 09:00 PM Central Time (US and Canada)

Join Zoom Meeting
https://us04web.zoom.us/j/79841255471?pwd=WjIyemduS2JKckZSaHBOUEM1Y21nUT09

Meeting ID: 798 4125 5471
Passcode: 5LEx5j

Zoom-Meeting.ics
701 bytes

(14) When it's time for the meeting to start, tap Meetings in the Zoom sidebar, then tap Start for your meeting.

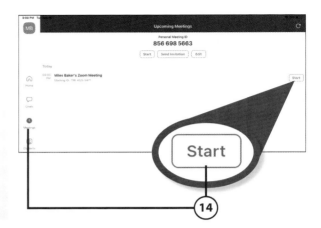

>>>*Go Further*
OTHER VIDEO CHAT SERVICES

FaceTime and Zoom are just two of many services that offer voice and video messaging and meetings. Other popular services include

- Facebook Messenger (www.messenger.com)

- Google Hangouts (hangouts.google.com)

- Google Meet (meet.google.com)

- Skype (www.skype.com)

Most of these services operate similarly to the ones discussed in this chapter and offer some level of free access. All have iPad apps available for download from the App Store. Don't be surprised to find a Google Meet or Skype invitation in your mailbox from someone you know!

Getting Social with Facebook, Pinterest, and Other Social Networks

Social networks are all the rage. I'm talking about online communities like Facebook, Pinterest, Twitter, and LinkedIn, places where you can go to share what you're doing with your online friends and find out what they're up to, too.

Social networking enables people to share experiences and opinions with each other via community-based websites. Whether you use Facebook, Pinterest, Twitter, LinkedIn, or some other social networking site, it's a great way to keep up-to-date on what your friends and family are doing—and you can do it from the comfort of your iPad. All you need to do is install the appropriate app(s) and learn how they work.

Using Facebook

With more than 2.7 billion active users each month worldwide, chances are many of your friends and family are already using Facebook.

People use Facebook to connect with current friends and family and reconnect with friends from the past. If you want to know what your friends from high school or the old neighborhood have been up to over the past several decades, chances are you can find them on Facebook.

In addition, Facebook helps you keep your friends informed about what you're doing. Write one post, and it's seen by hundreds of your online "friends." It's the easiest way I know to connect with almost everyone you know.

Like all the social networks discussed in this chapter, Facebook is completely free to use. You access Facebook from its iPad app, which you can download from Apple's App Store.

Navigate Facebook's iPad App

The first time you launch the Facebook app, you're prompted to either sign in to an existing account (if you have one) or create a new account. Follow the onscreen instructions to proceed from there.

Whenever you open the Facebook app, you see the News Feed screen. The News Feed is where you view status updates from all your Facebook friends, and the way it looks depends on how you're holding your iPad.

1. Tap the Facebook icon to launch the Facebook app.

2. In portrait mode (held vertically), you see the normal screen with no additional sidebars. All the navigation icons are at the bottom of the screen; tap News Feed to display the News Feed.

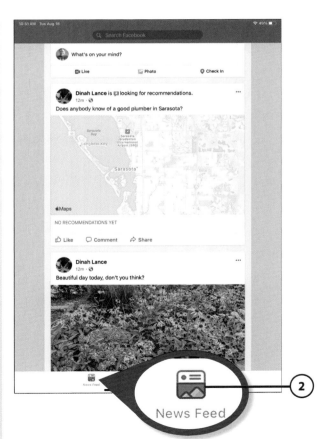

3 In landscape mode (held horizontally), you see the News Feed on the left with a sidebar on the right side of the screen that displays the Chat panel with its Search box, a list of recently active groups, and (sometimes) other items. Scroll down the screen to view more updates in the News Feed; refresh the News Feed by pulling down from the top of the screen and then releasing.

4 Tap the Friends icon to view new friends, friend requests, and people you might know.

5 Tap the Marketplace icon to see items for sale near you.

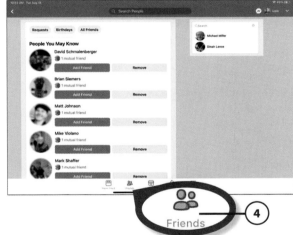

6 Tap the Notifications icon to view notifications from Facebook.

7 Tap the Menu icon to view your friends, events, groups, and more and to configure Facebook settings.

Pages and Groups

In addition to posts from individuals, Facebook offers official Pages from companies and celebrities, as well as topic-specific groups. You can "like" a public Page to receive posts from that Page, or join a group to post and receive messages from other members of that group.

Read and Respond to Posts

The messages people post to Facebook are called *status updates* or *posts*. You read and respond to posts from your friends on the News Feed screen.

1 For any given post, tap the Like icon to like that post. Tap and hold the Like icon to choose from a range of responses, from Love to Sad to Angry.

2 Tap the Comment icon to comment on a post.

3 Tap the Share icon to share this post.

4 Tap the poster's name to view that person's profile page, which displays her personal information, posts she's made, and photos she's uploaded.

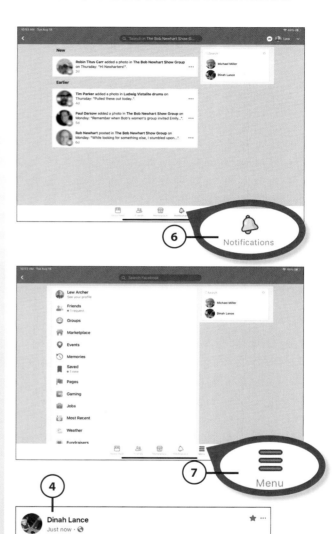

5 If the post includes a photo, tap the photo to view it full screen.

6 If the post includes a video, playback may start automatically, typically with the sound muted. Tap the video to view it (and hear it) on a separate video feed page.

7 If the post includes a link to another web page, tap the link or thumbnail to view that page on a new screen.

Post a Status Update

You create new status updates from the Facebook app's News Feed screen. The status updates you post are displayed on your friends' News Feeds.

(1) Tap within the status update ("What's on your mind?") box at the top of the page to display the Create Post panel.

(2) Type the text of your message into the big "What's on your mind?" text box.

(3) Change who can see the post by tapping the Privacy button beneath your name.

(4) Tap Photo/Video to include a photo or video with your post.

(5) Tap Tag Friends to tag another person in your post.

(6) Tap Feeling/Activity to share how you feel, tell others what you're doing, or add a small graphic sticker image to your message.

(7) Tap Check In to include your location in your post.

(8) Tap Post to post the status update.

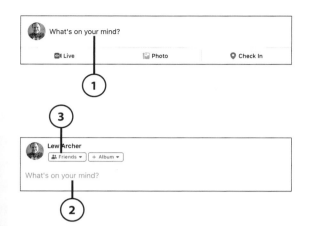

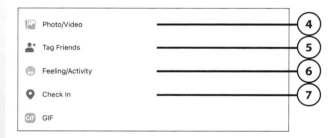

More About Facebook

Learn more about Facebook in my companion book, *My Facebook for Seniors*, available online and in bookstores everywhere.

Using Pinterest

Facebook isn't the only social network that may interest you. Pinterest is a newer, different type of social network with particular appeal to middle-aged and older women—although there are a growing number of male users, too.

Unlike Facebook, which lets you post text-based status updates, Pinterest is all about images. The site consists of a collection of virtual online boards that people use to share pictures they find interesting. Users "pin" or save photos and other images to their personal message boards, and then they share their pins with online friends.

You can save images of anything—clothing, furniture, recipes, do-it-yourself projects, and the like. Your Pinterest friends can then "repin" your images to their boards—and on and on.

Like Facebook, Pinterest is totally free to use. You access Pinterest from the Pinterest app.

View and Save Pins

You can download the Pinterest app for free from Apple's App Store. The first time you launch the app, you're prompted to either sign in to an existing account (if you have one) or create a new account. Follow the onscreen instructions to proceed from there.

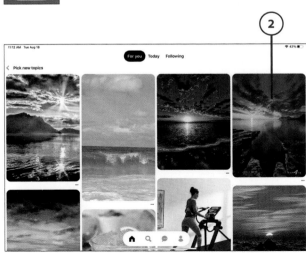

1. Tap the Pinterest icon to launch the Pinterest app. You see a variety of pins.

2. The pins in your feed include items pinned by people you follow, as well as recommended pins from Pinterest. Tap a pin to view it full screen.

3 Press and hold a pin to display the command icons.

4 Keep your finger pressed to the screen and move it to the Save icon to save this item to one of your boards.

5 Select which board you want to save to (or tap Create Board to pin to a new board). The item is then saved to that board.

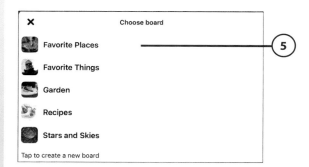

Not Always Welcome

Pinning images other than those you have personally created can infringe on copyright laws. Most websites don't object to people sharing their images on Pinterest, but some do and code their pages to prohibit pinning. If you try to pin from one of these pages, you get a message that no pinnable images have been found. If you pin an image that some entity owns and doesn't want you to pin, that site can ask Pinterest to take down the pin. If Pinterest removes one of your pins, it will notify you via email. Legally, Pinterest says it's not responsible for any copyright claims for items saved to its site. You, individually, could be responsible, although that has not been an issue to date.

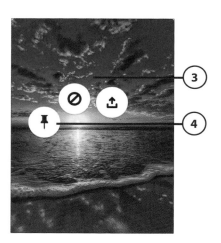

Pin from a Website

Many Pinterest users find images outside of Pinterest to pin to their boards. When you see an image on a website, you can easily pin it to a Pinterest board.

1 From the Safari web browser, navigate to the web page that contains the image you want to pin and then tap the Share icon in the browser to display the Share panel.

2 Scroll to and tap Pinterest to display the Pick an Image panel. (If you don't see the Pinterest icon, tap More to display it.)

3 You see thumbnails of all available images on that web page. Tap the thumbnail you want to pin.

4 Tap Next.

5 Tap the board you want to pin to (or tap Create Board to pin to a new board). The selected image is pinned to that board on Pinterest.

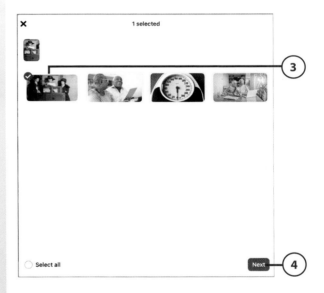

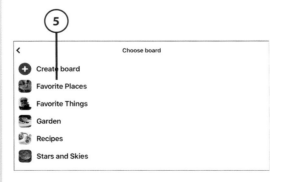

Using Twitter

You may have heard of Twitter, even if you haven't used it yourself. It's one of the more popular social media networks out there, although it isn't that widely used by those 50 and up. It's most popular among users in their twenties and thirties—and among politicians and celebrities reaching out to large audiences. Like other social networks, it is completely free to use.

Twitter is kind of like Facebook, but with only posts—called *tweets*—and no groups or pages or any of that. Tweets are short posts (280 characters or less), kind of like text messages, although they can include photos, videos, and links to pages on the Web.

As happens with the other social networks, you identify people you want to follow, and then you see all of their tweets in your feed. People who follow you see your tweets in their feed.

Read Tweets

Before you use Twitter, you need to sign up for an account, which you can do from the Twitter app, available for free from Apple's App Store. Once you do that, you're ready to tweet.

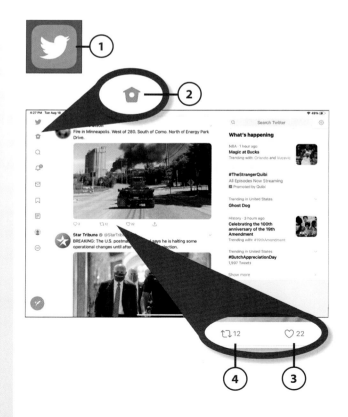

1. Tap the Twitter icon to open the Twitter app.

2. Tap the Home icon to display your feed. Tweets from people you follow are listed here.

3. To "like" a tweet, tap the Heart icon.

4. To retweet a tweet (that is, post it to the people who follow you), tap the Retweet icon, and then tap Retweet.

Create a New Tweet

Tweets can be a maximum of 280 characters long. You can include photos, videos, and web links in your tweets.

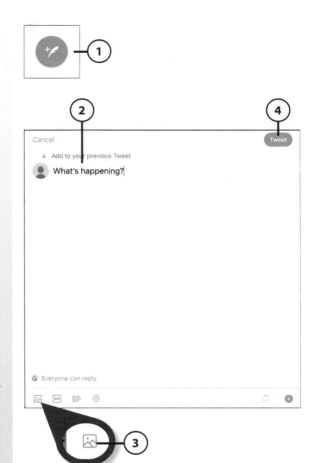

1. From the Twitter app, tap the New Tweet icon.

2. Use the onscreen keyboard to enter the text for your tweet. Remember to make it no more than 280 characters long!

3. To tweet a photo or video, tap the Photo icon and then select a photo or video stored on your phone. (The photo or video doesn't count toward the 280-character limit.)

4. Tap the Tweet button to post your tweet.

Hashtags

Twitter encourages the use of *hashtags* in tweets. A hashtag is a word or phrase (with no spaces between words) preceded by the hash (#) character, like this: #hashtag. Use hashtags to link your tweet to other tweets with the same hashtag; click a hashtag to see other tweets with that hashtag.

>>>Go Further
OTHER SOCIAL MEDIA

Facebook, Twitter, Pinterest, and LinkedIn may be the most popular social networks among older users, but they're not the only social media available today. There are several other social networks popular among other segments of the public, particular younger users.

Among the most popular of these other social networks are

- Instagram (www.instagram.com), designed for photo and video sharing, with more than 1 billion active users (primarily on their phones).

- Reddit (www.reddit.com), a combination news and discussion site with tons of "subreddits" dedicated to very specific topics.

- Snapchat (www.snapchat), targeting a younger demographic (average user age is 13) with photo and short video messages that "disappear" after a short period of time.

- TikTok (www.tiktok.com), a social network for short (60 seconds or less) and fun mobile videos; two-thirds of its users are under age 24 and a quarter are between 13 and 17 years old.

- WhatsApp (www.whatsapp.com), for secure text, voice, and video messaging.

All of these social media are free to use and have iPad apps available in the App Store.

Using LinkedIn

LinkedIn is a social network with a business bent. It's designed primarily for business professionals, and it's a good way to keep in touch with others in your profession, including people you've worked with in the past. LinkedIn is particularly useful if you're in the market for a new job; it's easy to make new contacts and do your business networking online.

Navigate the LinkedIn App

Like all the other social networks, LinkedIn is free to use; all you have to do is create an account and enter a little information about yourself. You access LinkedIn on your iPad from the LinkedIn app, which is available for download (for free) from Apple's App Store.

(1) Tap the LinkedIn icon to open the LinkedIn app.

(2) Tap Home to view posts from people or businesses you follow.

(3) Tap Like to like a post.

(4) Tap Comment to comment on a post.

(5) Tap My Network to connect with other users and view suggestions of people to follow.

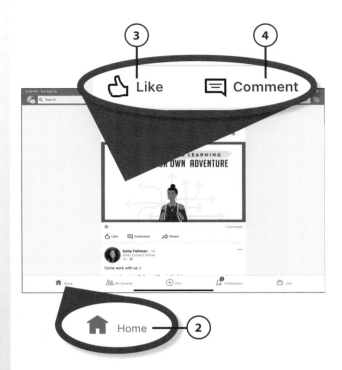

6 Tap Notifications to view recent activity from the people you follow.

7 Tap Jobs to look for employment opportunities.

6 Notifications

7 Jobs

Create a New Post

You can share notes, articles, photos, and more with the people you're linked to on LinkedIn.

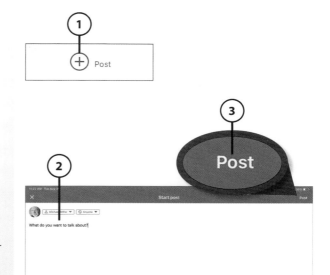

(**1**) Tap Post to display the Post screen.

(**2**) Enter the text of your message into the "What do you want to talk about?" text box.

(**3**) Tap Post to submit your post.

More About Social Media

Learn more about Facebook, Pinterest, Twitter, and other social networks in my companion book, *My Social Media for Seniors*, available online and in bookstores everywhere.

>>>Go Further
PRIVACY ON SOCIAL NETWORKS

It's important when using any social network to be aware of your privacy. It's easy to think of Facebook or any social network as a personal diary, but it's not. These social networks are extremely public; when you post a message or photo, it could be viewed by millions of people you don't even know.

If you value your privacy, you want to configure your settings on each network so that what you post is seen only by select people. Ideally, you want only your friends to see what you post; that means changing the posting privacy options from public to another more private setting. In addition, some networks (such as Facebook) let you adjust your privacy on a post-by-post basis. This way you could post something that you want your family members to see but don't want to show to co-workers or other friends.

How you adjust a network's privacy settings differs from network to network. In the Facebook app, for example, tap the Menu icon and then tap Settings & Privacy; everything you need to configure, privacy-wise, is listed there.

Of course, the best way to keep some things private on a social network is to not post them at all. You're old enough to know when to be discreet; resist the urge to post private information, private thoughts, and photos that ought to stay private. If you don't post 'em, nobody'll see 'em. If you post 'em, they exist in cyberspace forever.

Finally, never, ever post your private contact information on a social network—or anywhere online, for that matter. If you don't want strangers calling you up or showing up at your door, don't share your phone number or street address with them. In addition, never post about going out of town (or even out on the town); it's not unheard of for burglars to troll the social networks so they'll know when a house is empty and ripe for the looting.

Bottom line: Be careful out there!

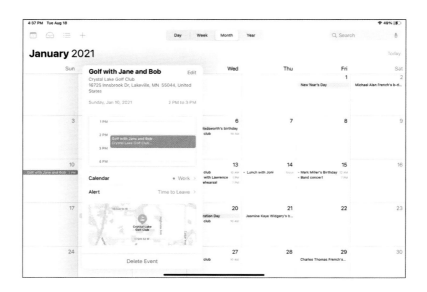

In this chapter, you learn how to use the iPad's Calendar and Reminders apps to manage your schedule and to-do lists.

→ Using the Calendar App
→ Using the Reminders App

13

Staying Organized with Calendar and Reminders

Sometimes it's tough to keep track of your time. How are you supposed to juggle all the various things you're supposed to do—appointments, events, and everything that's on your to-do list?

Fortunately, you can use your iPad to help you manage just about everything on your schedule. Use the Calendar app to manage your appointments and events and the Reminders app to track all the items on your to-do list.

Using the Calendar App

Apple's Calendar app is included with your iPad. Use the Calendar app to track everything you have on your schedule—what the app calls *events*.

View Events

You can view your calendar by day, week, month, or year.

(1) From the Home screen, tap the Calendar icon to open the Calendar app.

(2) Tap the Day tab to view a daily schedule. Scroll down to view times later in the day.

(3) Swipe the top of the calendar to the left to advance a week, or swipe the middle of the calendar to the left to view the next day.

(4) Swipe the top of the calendar to the right to view the previous week, or swipe the middle of the calendar to the right to view the previous day.

(5) Tap a specific date at the top of the calendar to view events for that day.

(6) Tap a specific event on the calendar to view details about that event.

(7) Tap Delete Event to delete the selected event.

(8) Tap the Week tab to view all events this week.

(9) Swipe the calendar to the left to view the next week.

(10) Swipe the calendar to the right to view the previous week.

(11) Tap an event to view details about that event.

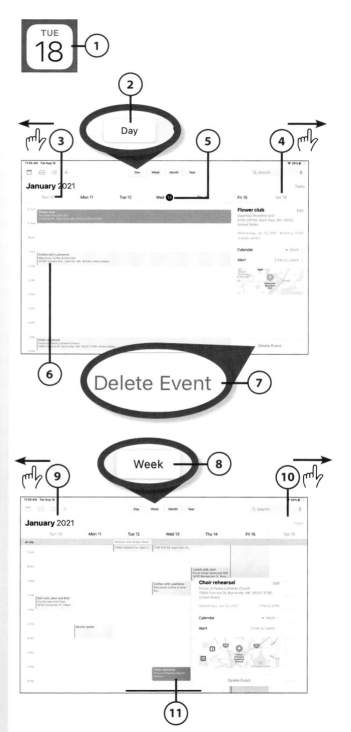

12 Tap the Month tab to view all events this month.

13 Scroll down the page to view events in the future.

14 Tap an event to view details about that event.

15 Tap the Year tab to view a yearly calendar.

16 Tap within a given month to view that monthly calendar.

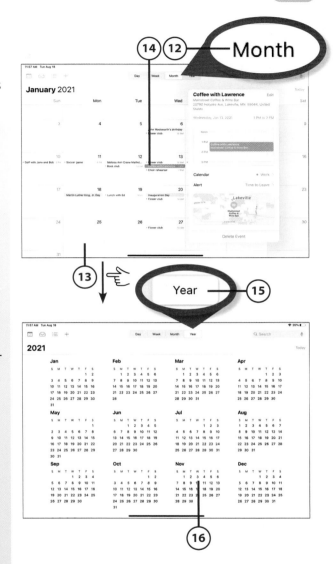

Add a New Event

By default, the Calendar app imports important dates from your contacts list—birthdays, anniversaries, and the like. It also creates events for major national holidays. (These show up in color shading.)

The real value of the Calendar app, of course, is evident when you add your own events to the calendar. You can add events for meetings, parties, doctor appointments, sports games—you name it.

1 From within the Calendar app, navigate to the day, week, month, or year you want and then tap the New (+) icon.

2 Tap within the Title field and add the name of this event.

3 Tap within the Location field and enter the location of this event. (This information is optional; the more detail you include—such as the street address—the better.)

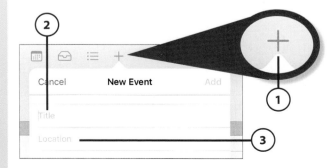

(4) If this event lasts all day, tap the All-Day switch. Otherwise...

(5) Tap the Starts field to expand the panel.

(6) Set the start time of the event.

(7) Tap the date for the event on the calendar.

(8) Tap the Ends field to select the end date and time for the event.

(9) If you need to travel to this event, tap Travel Time and add the appropriate travel time.

(10) Tap Add to add this event to your calendar.

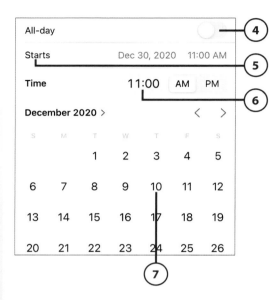

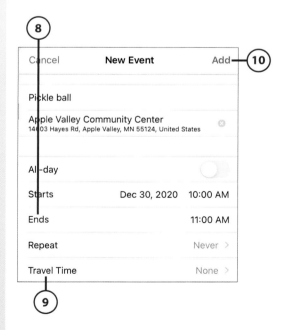

Create a Recurring Event

Some events repeat. For example, you might have a standing golf match every Saturday morning or a neighborhood meeting the first Monday of each month. Fortunately, the Calendar app lets you create recurring events.

1. From within the Calendar app, create a new event as normal and then tap the Repeat field.

2. Tap how often this event recurs—Every Day, Every Week, Every 2 Weeks, Every Month, or Every Year.

3. Tap Custom if the occurrence is more complex.

4. Tap Frequency to select how often the event recurs—Daily, Weekly, Monthly, or Yearly.

5. Tap Every to determine how the event recurs—every 1, 2, 3, or so days, weeks, months, or years.

6 If you selected Weekly frequency, select which day(s) of the week this event recurs.

7 If you selected Monthly frequency, select which day(s) of the month this event recurs. *Or…*

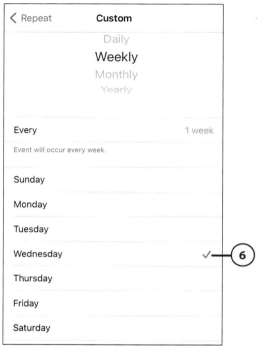

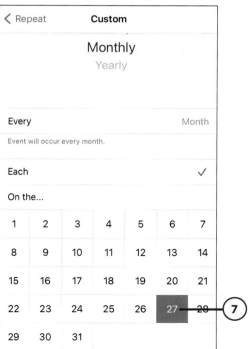

8 Tap On The and select which day(s) you want (First Sunday of the month, Second Thursday of the month, and so forth).

9 If you selected Yearly frequency (perfect for birthdays, anniversaries, and other annual celebrations), select which month(s) of the year and days of the week this event recurs.

10 Tap the back arrow to return to the Repeat panel, and tap again to return to the New Event panel.

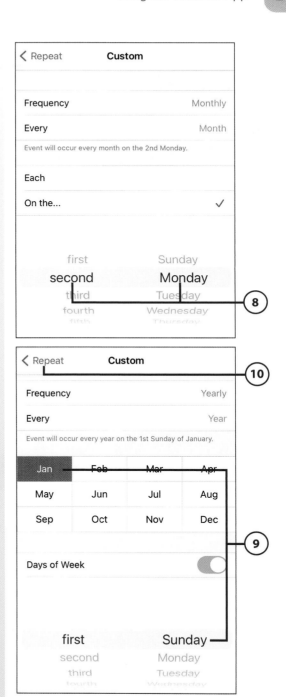

11 Tap the End Repeat field unless the event is ongoing. (If it's ongoing—that is, if it doesn't have a set end date—don't tap End Repeat, and skip steps 12 and 13.)

12 Tap On Date.

13 Tap the last date this event occurs.

14 Tap the back arrow to return to the New Event panel and finish creating the event. Tap Add when done.

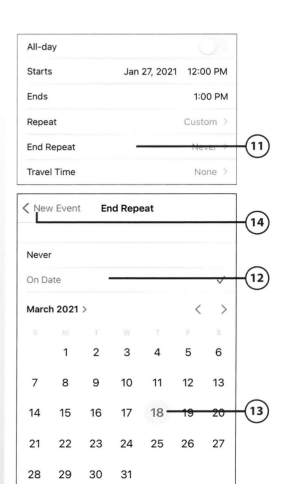

Create an Alert for an Event

Sometimes you want to be notified when an event is coming up. Maybe you need a full day to prepare for a big event or just a five-minute reminder to let you know what you need to do next. You can add alerts to any event you create in the Calendar app.

1 From within the Calendar app, create a new event as normal and then tap the Alert field.

(2) Tap when you want to be alerted, anywhere from at the time of the event to 1 week before. Your iPad displays an alert for this event at the designated time.

(3) If you've entered travel time for the event, you can be alerted when it's time to leave for the event (based on your location, the event's location, and current traffic conditions). Tap to select At Start of Travel Time.

Invite Others to an Event

Some events are personal. Others are quite public, involving lots of other people. (We're talking meetings, parties, and the like.) With the Calendar app, you can invite your friends and family to new events you create so they'll have these events on their calendars, too.

(1) From within the Calendar app, create a new event as normal and then tap the Invitees field.

(2) From within the Add Invitees panel, tap the To field and enter the name or email address of the person you want to invite.

(3) Alternatively, tap the + to display and select from people in your contacts list.

(4) This person is added to the invite list. Repeat steps 2 and 3 to add more invitees.

(5) Tap Done. Each invitee is sent an email inviting him or her to your event.

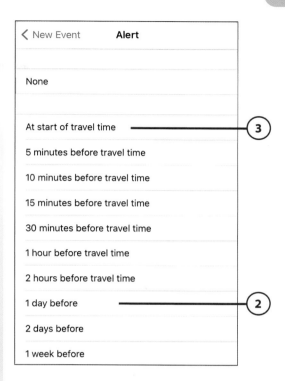

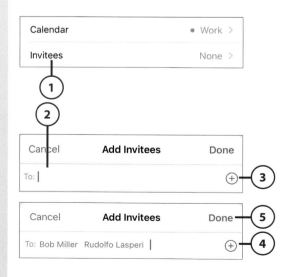

Using the Reminders App

I have trouble remembering things. Always have—it's not an age thing; I just have a poor memory. As such, I can use all the assistance I can get to help me track the various things I need to do.

With Apple's Reminders app, included with your iPad, you can create a digital to-do list containing all the various tasks and chores you need to remember. Add an item to the Reminders list and you'll be prompted (or nagged, as the case may be) to complete that task.

View and Manage Your Reminders

By default, upcoming and past-due reminders are displayed in the Reminders widget on your iPad's Cover Sheet screen (on your Lock screen). You also can review them all in the Reminders app itself.

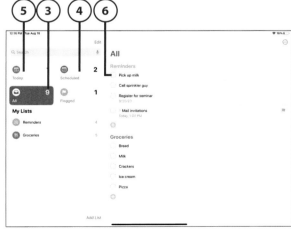

(1) From the Cover Sheet screen, tap a reminder to view it in the Reminders app. *Or…*

(2) From the Home screen, tap the Reminders icon to open the Reminders app.

(3) Tap All to view items you have not yet completed.

(4) Tap Scheduled to view items you want to complete by a given date.

(5) Tap Today to view those items due today.

(6) Tap the circle next to a given item to mark that item as completed.

Change the Order

To change the order of the items in your Reminders list, press and drag any item to a new position.

Add a New Reminder

Adding a new item to your reminders list is as easy as adding a new event to your calendar.

(1) From within the Reminders app, go to the My Lists section and tap Reminders.

(2) Tap + New Reminder.

(3) Enter the thing of which you need to be reminded.

(4) Tap Done. *Or…*

(5) Tap Information (i) to enter more information about this reminder.

(6) Tap "on" the Date switch to enter when you want to be reminded.

(7) Tap "on" the Time switch to enter a specific time when you want to be reminded.

(8) Tap "on" the Location switch to be reminded when you're near a specific location. (This is good for reminding you to pick up items at the grocery store.)

(9) Tap "on" the Messaging switch to be reminded when you're messaging with a specific person.

(10) Tap "on" the Flagged switch to mark an item as important.

(11) Tap Priority and then select Low, Medium, or High to set the priority for the item.

(12) Tap Done when done.

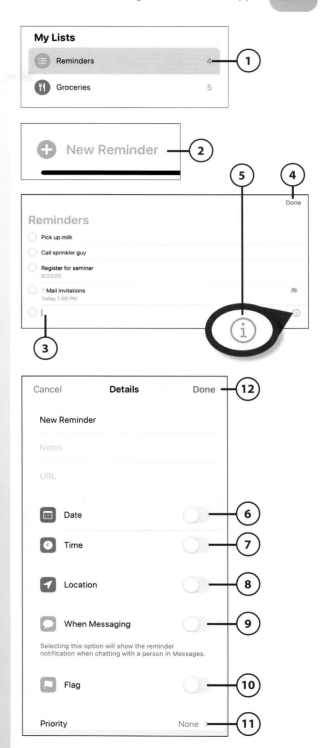

>>>Go Further

USE SIRI

Instead of entering events and reminders manually, you can use Siri to create new items for both the Calendar and Reminders apps.

To add a calendar event, activate Siri and speak as much information as you know about the event. Say something like, "Schedule a meeting Monday at 3:00 at the Mounds Conference Center" or "Create an event on Tuesday March 29th at 4:00 p.m."

To add a reminder, activate Siri and speak about the task. Say something like, "Remind me to pick up milk on the way home" or "Remember to pay the rent on the first of the month."

Adventure I

A Scandal in Bohemia

I.

To Sherlock Holmes she is always *the* woman. I have seldom heard him mention her under any other name. In his eyes she eclipses and predominates the whole of her sex. It was not that he felt any emotion akin to love for Irene Adler. All emotions, and that one particularly, were abhorrent to his cold, precise but admirably balanced mind. He was, I take it, the most perfect reasoning and observing machine that the world has seen, but as a lover he would have placed himself in a false position. He never spoke of the softer passions, save with a gibe and a sneer. They were admirable things for the observer—excellent for drawing the veil from men's motives and actions. But for the trained reasoner to admit such intrusions into his own delicate and finely adjusted temperament was to introduce a distracting factor which might throw a doubt upon all his mental results. Grit in a

sensitive instrument, or a crack in one of his own high-power lenses, would not be more disturbing than a strong emotion in a nature such as his. And yet there was but one woman to him, and that woman was the late Irene Adler, of dubious and questionable memory.

I had seen little of Holmes lately. My marriage had drifted us away from each other. My own complete happiness, and the home-centred interests which rise up around the man who first finds himself master of his own establishment, were sufficient to absorb all my attention, while Holmes, who loathed every form of society with his whole Bohemian soul, remained in our lodgings in Baker Street, buried among his old books, and alternating from week to week between cocaine and ambition, the drowsiness of the drug, and the fierce energy of his own keen nature. He was still, as ever, deeply attracted by the study of crime, and occupied his immense faculties and extraordinary powers of observation in following out those clues, and clearing up those mysteries which had been abandoned as hopeless by the official police. From time to time I heard some vague account of his doings: of his summons to Odessa in the case of the Trepoff murder, of his clearing up of the singular tragedy of the Atkinson brothers at Trincomalee, and finally of the mission which

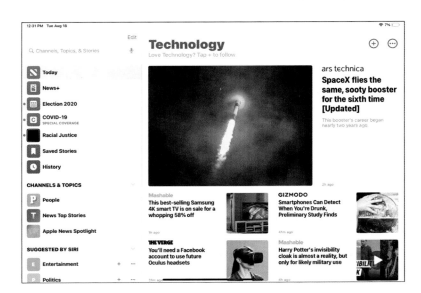

In this chapter, you learn how to use your iPad
to read various types of publications.

→ Reading eBooks on Your iPad
→ Reading News Stories with the News App

14

Reading eBooks, Magazines, and Newspapers

Your iPad is a great device for reading. You can use your iPad and the appropriate apps to read books, newspapers, magazines, and other articles—all in electronic format, all in the palms of your hands.

Reading eBooks on Your iPad

Electronic books, also known as eBooks, are a convenient way to read your favorite books. You can store multiple eBooks on your iPad and read any of them whenever and wherever you want. No more bulky paper books to haul around; no more losing your place among the dog-eared pages. Everything fits on and is viewed on your iPad.

The Apple Books app is preinstalled on your new iPad. You use the Books app to purchase eBooks from the Apple Book Store—and to read those eBooks and audiobooks you purchase.

Find and Purchase Books with the Books App

Before you can read eBooks with the Books app, you have to buy something to read—which you do directly from the Books app, from the Apple Book Store.

1. From the iPad's Home screen, tap the Books icon to open the Books app.

2. Tap Book Store to enter the Apple Book Store.

3. Tap Browse Sections to view all the categories in the Book Store.

4. On the Book Store home page, scroll to the New & Trending section to view new releases.

5. Tap See All to view all books in this section.

6. Scroll to the Top Charts section to view top books in various categories.

7 Scroll to the bottom of the page to view the most popular genres in the book store.

8 Tap All Genres to view all genres in the store.

9 Tap a genre to view books of that type. *Or…*

10 Tap Search to search by title, author, or subject matter.

11 Tap a book to view more details.

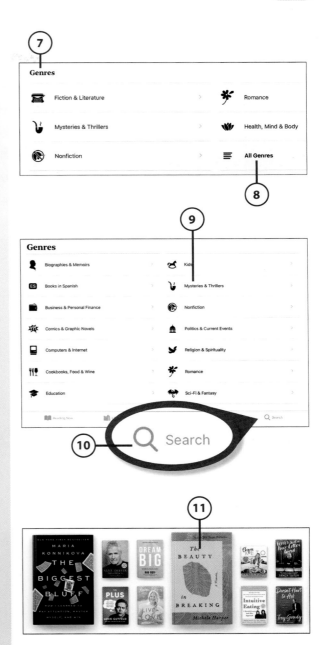

(12) Tap Sample to download a short sample of this book.

(13) Tap the Buy button to purchase the book. You see an Apple Books panel.

(14) Follow the instructions to confirm your purchase.

Amazon Kindle App

As an alternative to the Apple Book Store and Books app, consider the Amazon bookstore and Kindle app. You can purchase eBooks at www.amazon.com and then read them on the Amazon Kindle app, which you can download for free from Apple's App Store. The Kindle app works very much like Apple's Books app, but without the ability to purchase books within the app. You have to purchase books from Amazon using your web browser and then you can read them on the Kindle app.

Read a Book with the Books App

All the books you've purchased are downloaded to your iPad and displayed on the Library screen.

(1) From within the Books app, tap Library to view the books and samples you've downloaded to your iPad.

(2) Tap Collections to view your books by type.

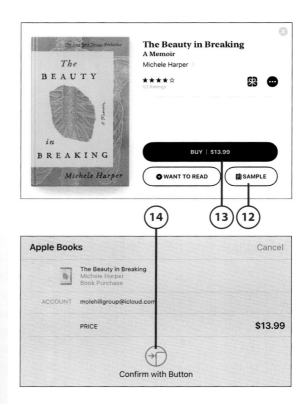

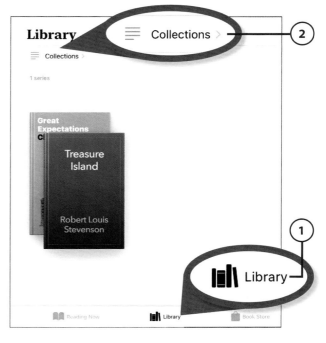

3 Tap Downloaded to view all books you've previously downloaded to your iPad.

4 Tap a book cover to open that book.

5 Swipe from right to left (or tap the right side of the screen) to turn to the next page.

6 Swipe from left to right (or tap the left side of the screen) to turn to the previous page.

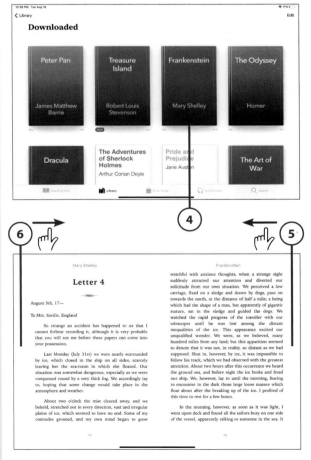

7 Tap the page to view additional controls at the top and bottom of the screen.

8 Drag the slider at the bottom of the page to move to another section in the book.

9 Tap the Fonts icon at the top of the screen to change the look and feel of the page.

10 Tap the right A to make the text larger and easier to read. Continue tapping to make the text even larger.

11 Tap the left A to make the text smaller. Continue tapping to make the text even smaller.

12 Tap Fonts to display and select a different font for the book's text.

13 Tap one of the colored circles to change the display theme for the book pages.

14 Tap "off" the Auto-Night Theme switch if you don't want the screen to change in low-light conditions. (Leave it on for better reading in bed at night.)

15 Tap "on" the Vertical Scrolling switch if you'd prefer to scroll through instead of flip through pages.

16 Tap the Search (magnifying glass) icon to display the Search box.

17 Type within the Search box to search for a given word, phrase, or page number.

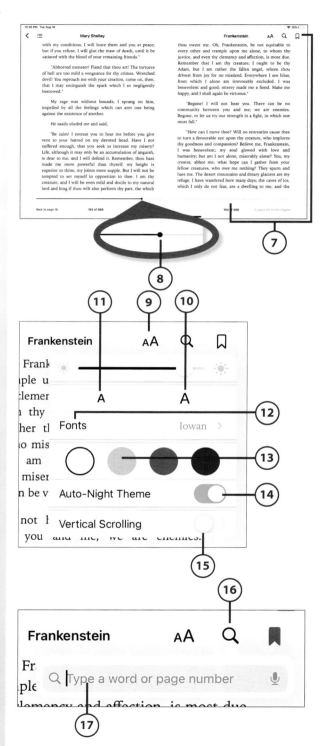

18 Tap the Bookmark icon to book-mark this page for future reference. (To view bookmarked pages, tap the Contents icon at the top left and select the Bookmarks tab.)

19 Tap the Contents icon to view the book's table of contents.

20 Tap the left-arrow icon to return to the Library screen.

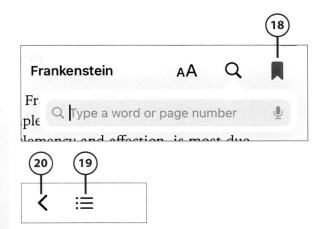

>>>Go Further

LISTENING TO AUDIOBOOKS

In addition to reading eBooks, you also can use the Books app to listen to audiobooks on your iPad. To purchase an audiobook, tap the Audiobooks icon to view all audiobooks for sale in the Apple Book Store. You can then purchase and download any audiobook just as you would an eBook.

To listen to an audiobook, tap that book on the Library page. Use the playback controls at the bottom of the screen to pause or resume playback, skip 15 seconds back or forward, and even change the speed of the playback.

If you've purchased audiobooks from Audible (www.audible.com), you must use the separate Audible app to listen to them on your iPad. The Audible app is available as a free download from Apple's App Store.

Free eBooks Online

Some eBooks are free. Amazon, Barnes & Noble, and the Apple Book Store offer some books for free download. If you're an Amazon Prime member, you get a variety of free eBooks as part of that subscription. Free eBooks are also available at sites such as Project Gutenberg (www.gutenberg.org) and Open Library (www.openlibrary.org). You can borrow eBooks from your local library, often via the Rakuten OverDrive website (www.overdrive.com). Ask your librarian for more information.

Reading News Stories with the News App

Books aren't the only things you can read on your iPad. You also can read online newspaper and magazine articles using Apple's News app.

The News app aggregates stories from a variety of news sources based on your reading habits. You see stories about topics you're most interested in from those sources you like the best.

Personalize What News You Receive

When you first launch the News app, you're prompted to select news sources (Apple calls them Channels) you'd like to include in your news feed. A handful of sources are already listed just in case you don't have any preferences yet. You can select additional news sources now or do so at any later time.

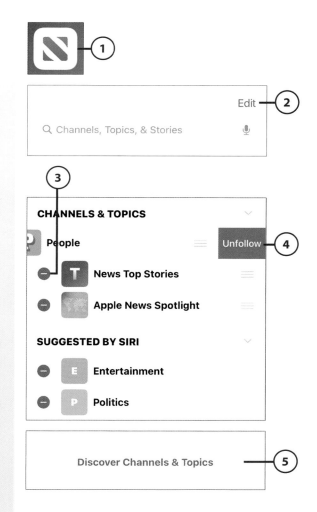

1. From the iPad's Home screen, tap the News icon to launch the News app.

2. Tap Edit at the top-left corner of the screen.

3. The left column displays suggested channels and topics currently appearing in your feed. Tap the red – button for any item you don't want in your feed.

4. Tap Unfollow or Ignore to remove that item from your feed; then tap Done.

5. Scroll to the bottom of the left column and tap Discover Channels & Topics.

6 You see a list of additional news sources you can add to your feed. Scroll through the thumbnails for suggested news sources, then tap the + to add any sources you want to follow.

7 Tap Done when you're done.

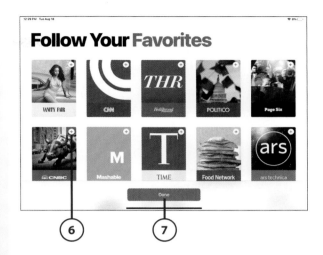

Read News Stories

The News app assembles stories based on the sources and topics you select and then fine-tunes those selections based on what you actually read over time. You see Top Stories first, then Trending Stories, then For You, and so on. Stories are displayed on the main For You page.

1 With the News app, stories are organized by news source and topic. Scroll down the screen to view additional sources and topics.

2 Tap a story to read it.

(3) Swipe up to read more of the story.

(4) Swipe right to left to go to the next story.

(5) Swipe left to right to go to the previous story.

(6) Tap the Font icon at the top of the page to change the size of the onscreen text.

(7) Tap the Share icon to share this story via instant message, email, or social media.

>>>Go Further
OTHER NEWS APPS

Apple's News app isn't the only way to get today's news on your iPad. You can use the Safari web browser to visit any news service's website, of course, and most major news sources also offer their own iPad apps for free download from the App Store.

Look for apps from ABC News, AP News (Associated Press), BBC, CBS News, CNN, Fox News, MSNBC, NBC News, the *New York Times*, NPR, Reuters News, *USA Today*, the *Wall Street Journal*, and the *Washington Post*. There are also several news aggregation apps that, like Apple News, gather news stories from a variety of sources. The most popular of these include Conservative News, Feedly, Flipboard, Google News, Microsoft News, and News Break.

In this chapter, you learn how to use your iPad to shoot photos and videos and then edit and share what you shot.

→ Shooting Digital Photos
→ Viewing and Editing Your Photos
→ Shooting and Editing Videos
→ Sharing Photos and Videos

Shooting, Editing, and Sharing Photos and Videos

Your iPad is much like an iPhone or other smartphone in terms of the features it offers (but with a much bigger screen, of course). This is especially true when it comes to photography because your iPad features two built-in cameras—one facing outward and one pointing toward your face. (Some iPad Pros have two cameras facing outward—the second for ultra-wide angle shots.) You can use these cameras to shoot still photos and videos and then use various apps to edit and share those items with others.

Shooting Digital Photos

Yes, you can use your iPad to shoot digital photos. You can take a picture of anything in front of you, or switch to the front-facing FaceTime camera and take a picture of yourself. (That's called a *selfie*.) You use the iPad's screen to preview the picture, and then aim the camera at whatever it is you're shooting and press the Shutter button. You can then edit the pictures you take to crop them, rotate them, adjust brightness and color, and even apply filters for special effects.

Launch the Camera App

Since you never know when a good photo op will present itself, Apple offers three different ways to quickly launch your iPad's camera.

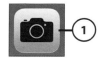

(1) From the iPad's Home screen, tap the Camera icon. *Or...*

(2) From any screen, swipe diagonally from the top-right corner to display the Control Center and then tap the Camera icon. *Or...*

(3) From the Lock or Cover Sheet screen, swipe from right to left. (That's right—you don't have to unlock your iPad to use the camera!)

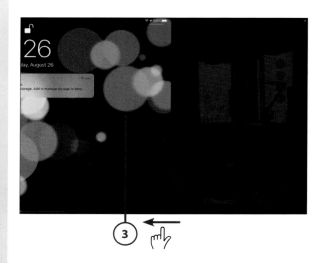

Take a Picture

When it comes to taking pictures with your iPad, you can shoot in the 16:9 aspect ratio that mirrors the iPad screen or in a square format better suited for some types of web uploads (such as your profile picture in Facebook). Use the rear-facing iSight camera to take pictures of things in front of you or (as discussed in the next task) the front-facing FaceTime camera to take selfies.

To take a portrait photo, hold your iPad vertically. To take a landscape photo, hold your iPad horizontally. The onscreen controls rotate accordingly.

1. Open the Camera app and make sure the rear-facing camera is selected. If not, tap the Switch button.

2. To use the timer (good if you want to include yourself in the shot—set the timer and run to get in front of the camera!), tap the Timer button and select either a 3- or 10-second timer. The timer activates when you press the Shutter button.

3. Tap Photo in the lower right to shoot 16:9 ratio photos.

4. Tap Square in the lower right to shoot square photos.

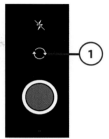

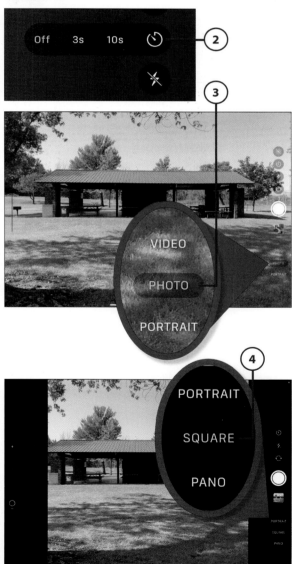

(**5**) Drag the zoom slider on the left side of the screen to zoom in to a shot. Alternatively, expand two fingers onscreen to zoom in.

(**6**) Tap the subject of the photo onscreen to focus on that person or object. This displays a box around the selected area and an Exposure slider; drag the slider up to increase the exposure (makes the shot brighter) or down to decrease the exposure (darkens the shot).

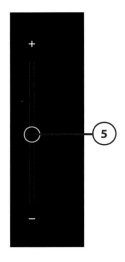

Auto Focus and Exposure

Your iPad's iSight camera includes auto focus and auto exposure, which uses face-recognition technology to identify the subject of the picture and focus and set exposure automatically. You can override these automatic controls by tapping to select a person or object onscreen different from the one your iPad selects.

(**7**) Use the iPad's display to aim the camera and compose your picture, then press the Shutter button to take the photo. Alternatively, press either Volume button on the side of your iPad to take the picture.

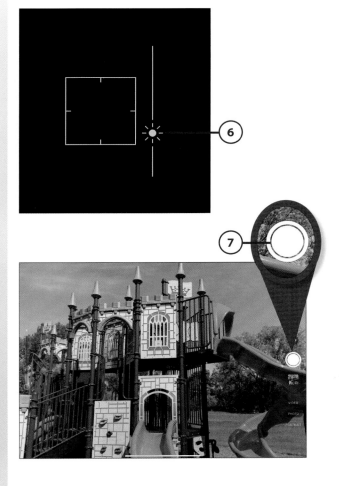

(8) Tap the preview thumbnail to view the picture you just took.

8

Burst Shots

To take a series of rapid-fire ("burst") shots, press and hold the Shutter button. Release the button to stop shooting.

>>>Go Further
SHOOTING ON AN iPAD PRO

If you have an iPad Pro, you have three other options when taking a picture.

Flash control

First, the iPad Pro has a built-in LED light on the back of the unit that functions as a camera flash to brighten dark scenes. Tap the Flash icon to select the desired flash setting: On, Auto, or Off.

Live photo icon

Second, there's a "live photo" option that records a short (1.5-second) video that starts just before you take a shot. Tap the Live Photo icon to turn on this option. (Live Photo isn't limited to iPad Pros; some newer regular iPads also have this feature.)

A Portrait mode selfie; note the blurred background

Finally, and perhaps most useful, iPad Pro models let you shoot with the front-facing camera in what's called Portrait mode. In this mode, you remain in focus while the background is blurred—perfect for selfies. You can use Portrait mode with the front-facing camera only—for selfie portraits of yourself. Select Portrait in the lower right to enter this mode.

The Depth or f-stop adjustment

When you shoot in Portrait mode, you see an "f" icon onscreen. Tap this to adjust the camera's f-stop, which affects depth of field. The lower the f-stop, the blurrier the background.

Portrait mode with lighting options

In Portrait mode, you also can select the type of lighting applied to the picture. This is done from a rotating carousel of options on the left side of the screen. You can choose from Natural Light, Studio Light, Contour Light, Stage Light, or Stage Light Mono (black and white).

Take a Selfie

Use the front-facing FaceTime camera on your iPad to take selfies of yourself. When you use the FaceTime camera, however, you have fewer shooting options available.

(**1**) Open the Camera app and tap the Switch button to switch to the front-facing FaceTime camera. Rotate the iPad vertically to take a portrait picture or horizontally to take a landscape photo.

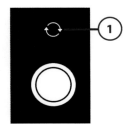

(2) Tap Photo to take a 16:9 ratio photo.

(3) Tap Square to take a square photo. (Great for profile pictures on many websites.)

4 On an iPad pro, tap Portrait to take a portrait photo with a blurred background.

5 Smile at the camera (and look at your smiling face onscreen) and then press the Shutter button to take the picture.

6 Tap the preview thumbnail to view the selfie you just took.

Screen Capture

To take a screenshot of what's currently displayed on your iPad's screen, simultaneously press and release the On/Off and Volume Up buttons (on an iPad Pro) or On/Off and Home buttons for other iPads. Screenshots are stored in the Photos app or iCloud Photo Library.

Take a Panoramic Photo

Apple's Camera app also lets you take panoramic photos that include a wide expanse not normally captured in a single photo. To create a panoramic photo, you shoot multiple photos from left to right, and the Camera app digitally stitches them together.

A panoramic photo

Panoramic photos are great for capturing outdoor landscapes, as well as "panning" a room indoors.

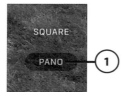

1. From within the Camera app, make sure the iSight (rear-facing) camera is selected, then tap Pano in the lower-right corner.

2. Hold the iPad vertically (portrait mode) and tap the Shutter button. (If you hold the iPad horizontally, the Photos app takes a vertical panoramic photo that stretches from the bottom to the top.)

3. As directed onscreen, move or pan the iPad slowly in the direction of the arrow. The progress of your pan appears in an inset over the main display.

(4) When you complete the pan, the camera returns to normal mode. (You can also stop the pan midway by tapping the Shutter button.) Tap the preview button to view your panoramic picture.

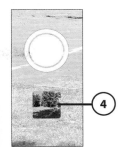

4

>>>*Go Further*

CONFIGURING THE CAMERA APP

Your iPad's Camera app works just fine "out of the box," as it were, but there are a few options you can configure to personalize the way it works. To configure the Camera app, tap the Settings icon and then select Camera in the left column.

Here are a few of the settings you might want to change:

- To have the Camera app remember the last type of picture you took (Photo, Video, or Square), tap Preserve Settings and then tap "on" the Camera Mode switch.

- If you're shooting on an iPad with Live Photo (which shoots a short video before every picture you take), you can turn it off by tapping Preserve Settings and then tapping "off" the Live Photo switch.

- Tap "on" the Grid switch to display an onscreen grid to help you better compose your photos.

- By default, the Camera app can shoot and read the QR codes you see on many products and items. To turn off this feature, tap "off" the Scan QR Codes switch.

- By default, the iSight camera records video at 1080p HD resolution with a frame rate of 30 fps. To shoot at a higher (4K) or lower (720p) resolution at a higher or lower frame rate, tap Record Video and make a new selection. (The available options depend on which model iPad you have; iPad Pro models have better cameras than regular iPads.)

- Some iPad models shoot in what is called "smart" HDR. (HDR stands for *high dynamic range*, and combines images shot at different exposures for better contrast.) This setting is enabled by default; to disable it, tap "off" the Smart HDR switch. And to keep the normal photos used to create the HDR photos, tap "on" the Keep Normal Photo switch. (Note that some iPads have manual HDR that you can enable from the Camera app's shooting screen by tapping "on" the HDR icon.)

The default settings work fine in most cases, but if you need to change any of them, that's how to do it.

Viewing and Editing Your Photos

Although the iPad may be a little inconvenient as a day-to-day digital camera, it's actually pretty good for viewing and editing photos. The large screen, combined with handy touchscreen operation, makes it easy to look at the photos you've taken and make minor changes to them if necessary.

You use Apple's Photos app to both view and edit your photos. In fact, you can use the iPad's Photos app to view and edit not only pictures taken with your iPad but also photos taken by any of your Apple devices. When you use the iCloud Photo Library, all photos you take with your iPad and iPhone are automatically uploaded and stored online, and available to all your Apple devices. So take a picture with your iPhone and edit it with your iPad—it's the best way to use both devices!

>>>*Go Further*

CONFIGURING THE PHOTOS APP

Just as you might want to configure a few settings in your iPad's Camera app, you also might want to customize a handful of settings in the accompanying Photos app. Go to your iPad's Settings page and tap Photos in the left column, and then look at the following settings in the right column:

- Tap "on" the iCloud Photos control to automatically upload your pictures to your iCloud account.

- To save storage space on your iPad, select Optimize iPad Storage; full-resolution photos will still be stored online in your iCloud account.

- If you'd rather keep full-resolution versions on your iPad, select Download and Keep Originals instead.

There are more settings you can configure, but these are the main ones you want to pay attention to.

View Your Photos

You can view photos you've just taken from within the Camera app or from Apple's Photos app.

(1) From within the Camera app, tap the preview thumbnail to display the photo full screen. (Tap the back arrow to return to the camera to shoot some more photos.) *Or…*

(2) From the Home screen, tap the Photos icon to open the Photos app.

(3) You can let the Photos app organize your photos automatically by date taken. You can also organize them manually in albums. To view the photos by date, tap Library.

(4) Tap Years to see photos organized by year taken.

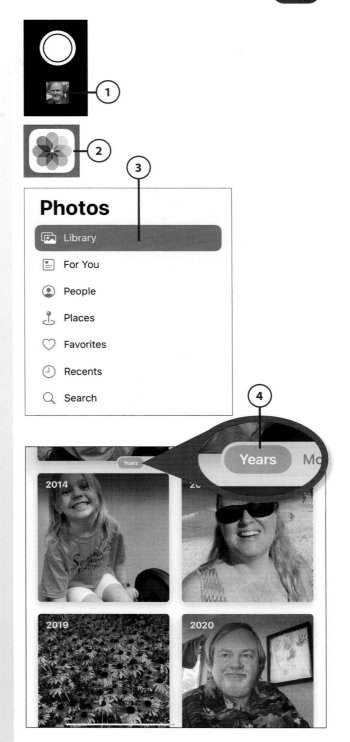

(5) Within a given year's photos, the Photos app organizes photos by month. You can tap a month within the Year grouping, or just tap Months in the floating menu.

(6) Within a month's photos, pictures are organized by specific day. You can tap a date after opening the month grouping, or just tap Days in the floating menu.

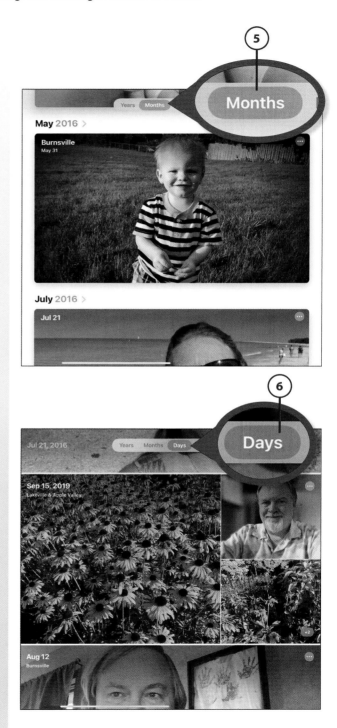

7 Tap All Photos to view all photos in chronological order.

8 Tap For You to view featured photos and shared album activities.

9 Tap People to view photos organized by person in the photo, identified via facial-recognition technology.

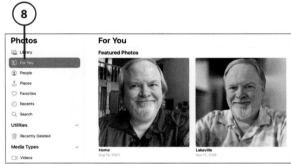

10 Tap Places to view a map of the United States with picture locations noted on the map.

11 Tap Favorites to view photos you've tagged as favorites.

12 Tap Recents to view recently taken photos

13 Tap Search to search for photos by keyword—people, places, categories, and so forth.

(14) To search for photos by media type, go to the Media Type section. What you see here depends on the types of photos you've taken.

(15) Go to the My Albums sections to view your photos organized by album.

(16) Tap an album to view the photos stored within.

(17) Tap to view a specific photo.

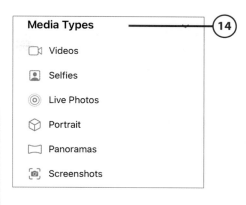

18 Tap the Heart icon to add this photo to your Favorites album.

19 Tap the Trash Can icon to delete the photo.

20 Additional photos in this group or album are displayed as thumb-nails in a strip along the bottom of the screen. Drag the strip left or right to see additional photos. (You can also swipe left or right on the main photo to see other photos in the group or album.)

21 Tap the Back icon to return to the previous group or album.

>>>Go Further

USING THE iCLOUD PHOTO LIBRARY

Although you can store your photos locally on your iPad, you can store many more photos online in Apple's iCloud Photo Library. When you enable the iCloud Photo Library on your device, all photos and videos you shoot are automatically uploaded to and stored on Apple's iCloud online storage service.

You can then access your photos and videos from any connected computer or device just by going to the iCloud website (www.icloud.com) and logging in with your Apple ID. Any edits or changes you make to the stored photos are visible on all your devices.

To enable iCloud Photo Library storage, tap the Settings icon, select Photos, and then tap "on" the iCloud Photos switch. You're ready to go.

Organize Photos in Albums

The Photos app does a good job automatically organizing your photos. That said, you might want to organize your photos your own way by creating virtual photo albums to group photos by type, activity, location, and so forth.

(1) From within the Photos app, tap All Albums to display your existing albums. (If you're within a given album, tap the Back icon to return to the main Albums screen.)

(2) Tap an album to view the photos stored within. *Or...*

(3) To create a new album, tap the + icon.

(4) Tap New Album.

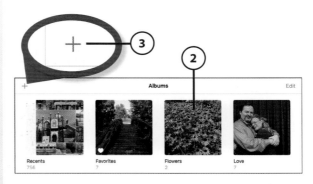

(5) From the New Album panel, enter a name for this album.

(6) Tap Save.

(7) You're prompted to add photos to your new album. Tap to select those photos you want to add. (You can select multiple photos; as you select each photo, a blue check mark appears in the corner.)

(8) Tap Done. The selected photos are added to your new album.

(9) To add additional photos to an album, open the group or album that contains the photo(s) you want to add and then tap the photo(s) you want.

(10) Tap the Share icon.

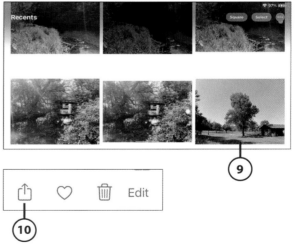

(11) Tap Add to Album.

(12) Tap to select the album to which you want to add these pictures. The pictures are now added.

Slideshows

The Photos app enables you to play a slideshow of photos from any given album. Just open the album, tap the More (…) icon, and then tap Slideshow.

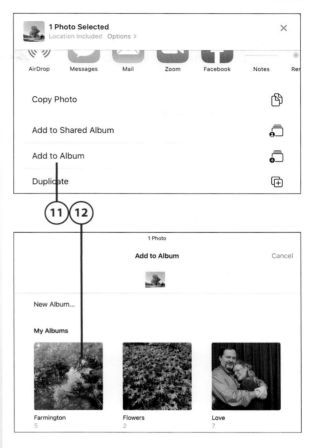

Crop and Straighten a Picture

Not every picture you take is perfect. Fortunately, digital pictures can be easily edited. You use the Photos app to make any necessary corrections to your photos.

(1) From within the Photos app, tap to open the photo you want to edit.

(2) Tap Edit to display the editing screen.

(3) Tap the Crop button.

(4) Crop the picture by dragging the corners of the screen until you like what you see.

(5) Rotate the picture 90 degrees clockwise by tapping the Rotate button at the top of the screen.

(6) Straighten or rotate the picture in smaller increments by dragging your finger up or down on the rotation control until the picture is as you like it.

(7) Tap Done when you're satisfied with the results. (If you don't like the changes, tap Cancel in the top-left corner to undo them and return to the original photo.)

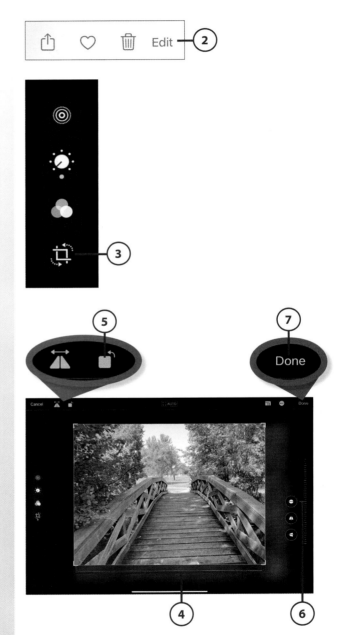

Apply Auto-Enhance

Sometimes a picture doesn't look quite right, but you don't want to do a lot of manual fine-tuning to fix it. In these instances, use the Photo app's Auto-Enhance control and let the app do the fixing for you. Auto-Enhance adjusts brightness, contrast, color, and other controls to try to make your picture look better.

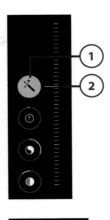

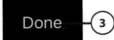

(1) From within the editing screen for a photo, tap the Auto-Enhance (magic wand) button.

(2) Drag the slider up or down to apply less or more of the enhancement.

(3) If you like what you see, tap Done. (If you don't like the results, tap the Auto-Enhance button again to turn off the auto enhancement.)

Apply a Filter

Another easy way to change the way a picture looks is to apply a photo filter. The Photos app comes with more than a half-dozen filters that can change the mood of a picture.

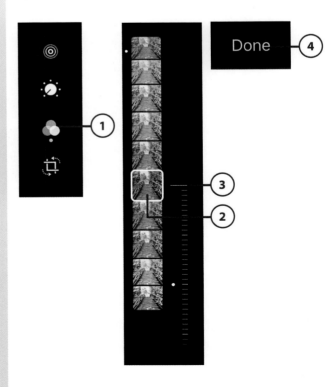

(1) From within the editing screen for a photo, tap the Filters button.

(2) You see the available filters. Tap a filter to apply it to your photo.

(3) Use the slider to adjust the intensity of the filter.

(4) If you like this filter, tap Done. (If you don't like this filter, select another or tap Cancel.)

Adjust Brightness, Color, and More

The Photos app also offers more traditional editing controls you can use to enhance your photos—exposure, brightness, contrast, saturation, and more.

1. From within the editing screen for a photo, tap the Adjustments button.

2. On the right, tap the control you want to adjust. You can choose from (top to bottom): Exposure, Brilliance, Highlights, Shadows, Contrast, Brightness, Black Point, Saturation, Vibrance, Warmth, Tint, Sharpness, Definition, Noise Reduction, and Vignette.

3. Use the slider to the side of the selected control to adjust that particular setting.

4. Tap Done when you're done editing. (Or tap Cancel to undo your changes and return to the original photo.)

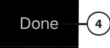

Shooting and Editing Videos

You can shoot videos with the rear-facing and front-facing cameras in your iPad. That means you can shoot videos of family gatherings, vacations, sporting events, and more—even selfie videos of yourself!

As with shooting still pictures, hold your iPad vertically to take a portrait video or horizontally to take a landscape video. The onscreen controls rotate accordingly.

Widescreen Videos

You should shoot your videos in landscape format, which requires you to hold your iPad horizontally. This is the orientation used by all movies and TV shows; it produces a 16:9 ratio widescreen picture, which fits perfectly on your TV screen.

Shoot a Video

Shooting a video with your iPad is almost as easy as shooting a still photograph.

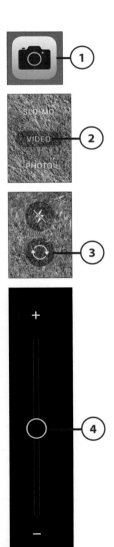

1. Tap the Camera icon to open the Camera app.

2. Tap Video in the lower-right corner of the screen.

3. Tap the Switch button to switch between the rear-facing (iSight) and front-facing (FaceTime) cameras.

4. Drag the zoom slider on the left side of the screen to zoom in to a shot. Alternatively, expand two fingers onscreen to zoom in.

(5) Optionally, tap the subject of the video onscreen to focus on that person or object. This displays a box around the selected area and an Exposure slider; drag the slider up to increase the exposure (makes the video brighter) or down to decrease the exposure (darkens the video).

(6) Use the iPad's display to aim the camera and compose your picture and then press the red Record button to start recording. Alternatively, press either Volume button on the side of your iPad to initiate recording.

(7) As you record, the elapsed time is displayed at the top of the screen.

(8) Tap the Record button again (or press either Volume button) to stop recording.

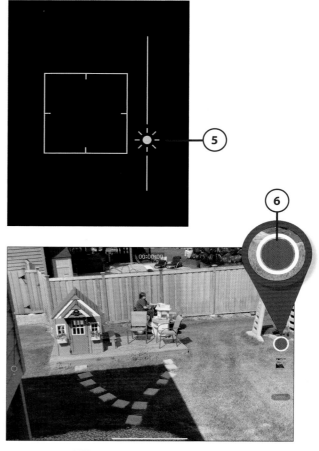

>>>Go Further

SLO-MO AND TIME-LAPSE VIDEOS

The Photos app enables you to take two special types of videos. You can shoot video for slow-motion playback or record a series of still photographs as a time-lapse video.

To shoot in slow motion, select Slo-Mo as the recording type. Your video is then shot at a super-fast 120 frames per second (fps). When a fast fps video is played at a normal (30 fps) frame rate, you get crystal-clear slow motion. You select which parts of your video to play in slow motion when you edit the video.

To record a time-lapse video, you have to keep your iPad perfectly still over the period of time you're shooting. Select Time-Lapse as the recording type, tap the Record button, and the Camera app starts capturing individual shots. Tap Record again to end the recording, and the individual shots are compiled into a short video.

Trim a Video

After you've shot a video, you can trim the beginning and end to make the video shorter and focus on the important parts. You can do this editing from within the Camera or Photos apps.

1. After you've shot the video in the Camera app, tap the thumbnail to display the video playback screen. *Or…*

2. From within the Photos app, tap to open the video you want to trim.

3. Tap Edit.

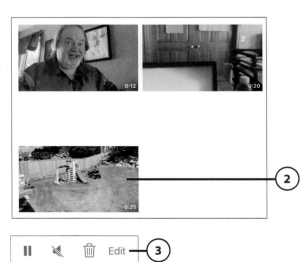

(**4**) In the series of thumbnails (called a *timeline*) beneath the video, touch and hold either the left or right arrow until a yellow border appears in the timeline, and then drag the yellow arrows until you've selected the portion of the video you want to keep. (Everything outside the yellow selected section will be trimmed.)

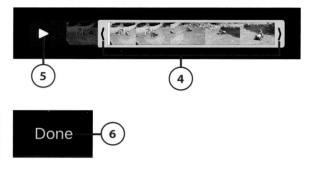

(**5**) Tap the Play button to play the selected (yellow) section of the video.

(**6**) Tap Done when you're done.

Other Editing Controls

All of the editing controls available for photographs are also available for editing your videos, including filters, cropping, exposure, you name it. Just open the video for editing, choose the options you want to adjust, and have at it! Tap the Play button at the bottom of the screen to see the results of your edits, and tap Done when you're done.

Sharing Photos and Videos

When you take a photo or video that you like, you might want to share it with friends and family. Fortunately, the Camera and Photos apps both make it easy to share your pics and vids via text message, email, social media, and more.

Share via Text Message

You can share photos and videos with your contacts via the Messages app.

1 From within the Camera or Photos app, open the item you want to share and then tap the Share icon.

2 Tap Messages to display the New Message pane.

3 Enter the name of the person you want to share with into the To field. (You can enter multiple people if you want.)

4 The photo or video is added to the conversation. Tap within the item to add a text message.

5 Tap the Send icon to send the message.

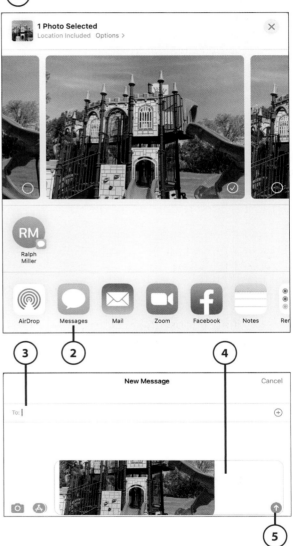

Share via Email

You can send a photo or video to any of your contacts via email using your iPad's Mail app.

(1) From within the Camera or Photos app, open the item you want to share and then tap the Share icon.

(2) Tap Mail to display the New Message pane.

(3) Tap within the To field and enter the names or email addresses of the intended recipients.

(4) Tap within the Subject field and enter a subject for this email.

(5) Tap within the main text area and enter any accompanying message.

(6) Tap Send to send the email.

Share via Facebook

If you're on Facebook or other social media, you can post a photo or video to that social network. This example uses Facebook, but the steps are similar if you're posting to Twitter or other networks.

1. From within the Camera or Photos app, open the item you want to share and then tap the Share icon.

2. Tap Facebook to display the Facebook pane.

3. Tap within the Say Something About This Photo field and enter the text of your post.

4. Tap to select your privacy options—who you want to see this post.

5. Tap Next.

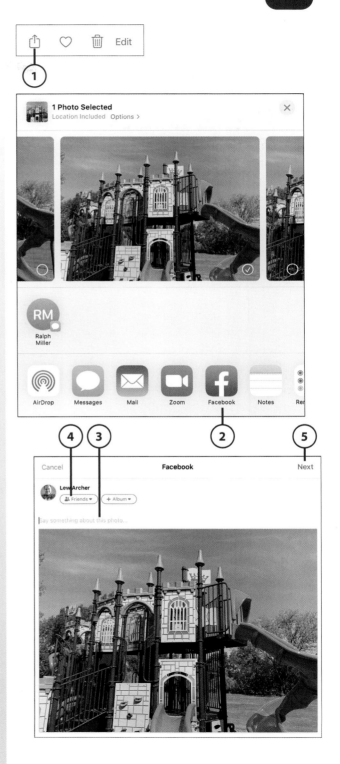

(6) Check how you want to share: News Feed, Your Story, Album, Friend's Timeline, or Group.

(7) Tap Share to post this photo or video as a new status update.

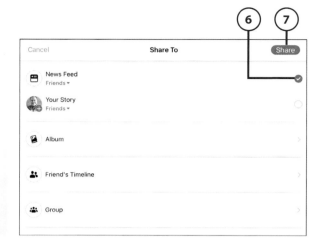

>>>Go Further
SHARING PHOTO ALBUMS

If you've enabled iCloud photo sharing and created shared folders for your friends and family members, you can upload your photos and videos to a shared folder. Just open the photo or video, tap the Share button, and then tap Add to Shared Album. You can add a comment to accompany this item and select which shared album you want to post to.

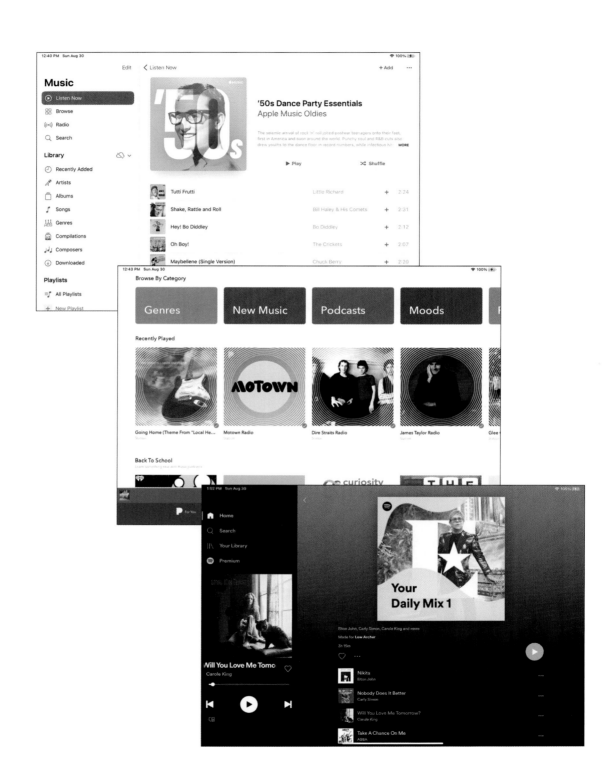

In this chapter, you learn how to listen to music on your iPad.

→ Listening to Streaming Music
→ Buying and Playing Music from the iTunes Store

16

Listening to Music

If you like to listen to music, you're in luck. Your iPad is a versatile music playback system that enables you to listen to the music you love wherever you happen to be. You can purchase and download music directly to your iPad or stream just about any song in the world over your iPad if you have an Internet connection.

Listening to Streaming Music

The way we listen to music has changed over the years. When I was growing up in the 1960s, I could listen to whatever was playing on AM or FM radio or buy LPs and singles to play on my RCA portable record player. (Mercifully, I kind of skipped right over the 8-track era.) Later generations listened to audiocassettes on their portable Walkman devices or purchased albums on digital compact discs (CDs). Still later generations learned how to download music from the Internet, either legally (through the iTunes Store) or illegally (via the infamous Napster).

That's all old news, however, as digital downloads are being supplanted by the concept of streaming music services. These services, such as Apple Music, Pandora, and Spotify, stream huge music libraries (tens of millions of tracks) over the Internet direct to any connected device—including your iPad. All you need is an Internet connection, and just about any song you want is available for your listening pleasure. You may need to create an account or purchase a subscription, but you don't need to buy or download individual tracks to your device.

Free—or Not

Most streaming music services offer some type of free listening level, typically supported by commercials. If you want to get rid of the commercials (and gain more control over what you listen to), paid subscriptions are available.

Listen to Apple Music

Apple offers its own streaming music service with more than 60 million individual tracks. Not surprisingly, it works well with your iPad.

Unlike most other streaming services, Apple Music doesn't have a free level; you pay $9.99 per month for an individual membership after a free three-month trial. (Other pricing is also available, including a $14.99 per month family membership.)

You access Apple Music from the Music app on your iPad. To sign up for the service, tap the For You tab. When prompted, opt to get your three months free, choose which plan you want (Individual, Student, or Family), and proceed from there.

When you first log on (and provide your Apple ID and password, of course), you're prompted to choose your favorite music, which you do by selecting the genres and artists you like best. Apple Music then fills up the For You tab with music it thinks you might like.

1. From the iPad Home screen, tap the Music icon to launch the Music app.

2. Tap Listen Now to view recommended playlists, albums, and artists.

3 Tap Browse to browse through new music.

4 Tap Radio to listen to Apple Music's preprogrammed radio stations.

5 Tap Search to search for specific music.

6 Enter your search term and then tap the result you want.

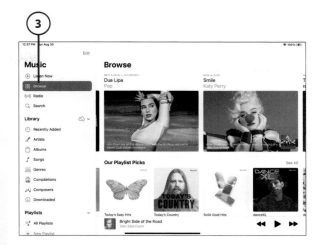

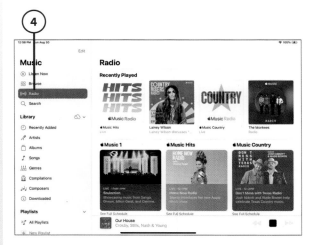

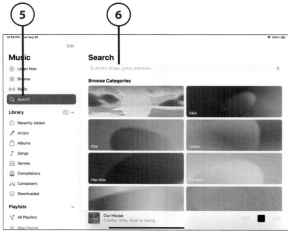

(7) Tap to listen to a given track, station, album, or playlist.

(8) Tap Play and you see playback controls in the bottom-right corner of the screen. Tap Pause to pause playback; tap Play to resume playback.

(9) Tap Next to play the next track.

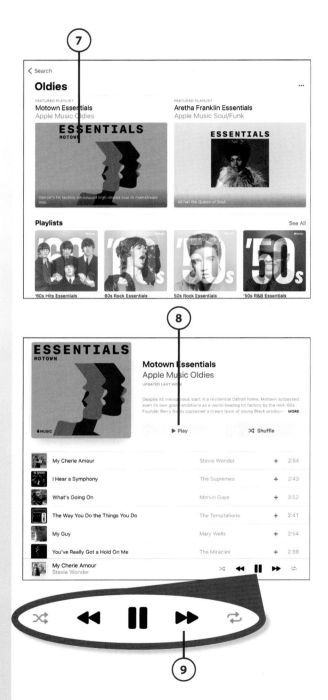

Listen to Pandora

Pandora is another popular streaming service, available on all platforms—not just on Apple devices. You listen to Pandora via the Pandora app, available for free from Apple's App Store.

Pandora offers three different subscription plans, each with unique features:

- **Basic:** Pandora's basic plan is free, although you need to subscribe and sign in. The free plan is much like traditional AM or FM radio; you listen to the songs Pandora selects for you, along with accompanying commercials. It's a little more personalized than traditional radio, however, in that you create personalized stations (up to 100). All you have to do is choose a song or artist; Pandora then creates a station with other songs like the one you picked. This plan is ad-supported, so you'll be subjected to commercials every handful of songs—just like traditional radio.

- **Pandora Plus:** If you want to get rid of the ads, go with the Pandora Plus plan. It's just like the free plan but without the commercials. For this, you pay $4.99 per month.

- **Pandora Premium:** Pandora's high-end plan, at $9.99 per month (or $14.99 per month for the Family plan that includes up to six accounts), is a bit different. In addition to the personalized stations, you also get on-demand playback. That is, you can search for and play any song in the Pandora library. You also can create personalized playlists of songs.

When you first launch the Pandora app, you're prompted to either sign up or log in, so do one or the other.

1. From the iPad Home screen, tap the Pandora icon to launch the Pandora app.

2 If it's not already selected, tap the My Collection screen to display all your existing radio stations.

3 Tap a station to listen.

4 Tap the Pause button to pause playback. The button turns into a Play button; tap Play to resume playback.

5 Tap the Thumbs Up icon if you like a track; the radio station will be fine-tuned to play more tracks like this.

6 Tap the Thumbs Down icon if you don't like a track; Pandora immediately skips to the next track and fine-tunes the radio station to not play music like the track you dislike. (You only have a limited number of skips available with the free subscription.)

7 Tap the Next button if the track is okay but you just don't want to listen to it now. The track will be skipped, but it won't affect how Pandora fine-tunes your station.

8 Tap the Replay button to play this track again. (You may not be able to do this if you have the free version.)

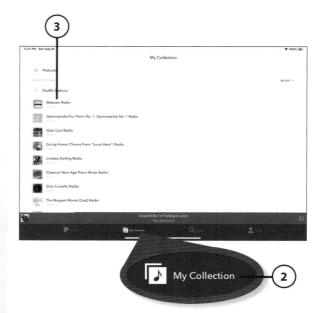

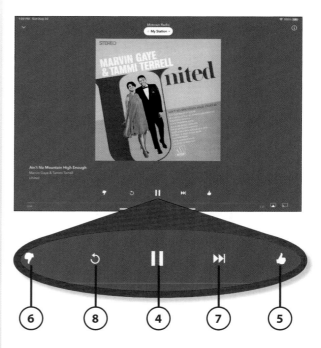

9 Find a new station by tapping the back arrow to return to the previous screen and then tap For You.

10 Swipe up and down to browse through genres, decades, top stations, recommended artists, and more.

11 Tap to listen to a station.

12 Tap Search to search for a particular song, artist, or genre.

13 Use the onscreen keyboard to type in the name of a song, artist, or genre.

14 Pandora lists matching artists, songs, stations, and more. Tap a selection to create that station, add it to your station list, and begin playback.

Listen to Spotify

Spotify is another big streaming music service. It works like Apple Music in that you can "dial up" the specific artists and songs you want to listen to. Spotify has more than 40 million songs in its online library.

You can choose from a free version that works more like Pandora, by throwing a lot of stuff you may not want into the "stations" you create. You're also subjected to commercials every few songs. If you want to get rid of the commercials and get unlimited skips, sign up for the $9.99 per month Spotify Premium plan (or $14.99 per month for the Family plan, which includes up to six accounts).

1. From the iPad Home screen, tap the Spotify icon to launch the Spotify app.

2. Tap to open the Home tab.

3. Scroll through the various sections, such as Recently Played, Trending Now, Popular New Releases, and Mood. Tap an item to start listening.

4. Tap Search to search or browse for specific tracks or artists.

5. Enter a query into the search box. Or…

6. Tap a genre or type of music to listen to.

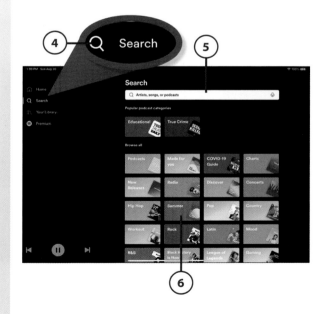

(7) Tap to listen to a playlist.

(8) Tap a specific track to listen to that song. *Or…*

(9) Tap Play to play all the songs in the playlist.

(10) When the music starts playing, you see the playback controls at the bottom left. Tap the Play/Pause button to pause playback.

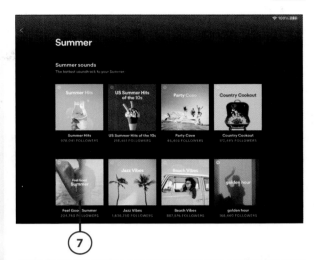

(7)

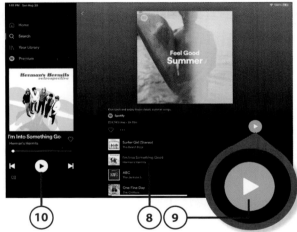

(10)   (8) (9)

>>>Go Further

OTHER STREAMING MUSIC SERVICES

Apple Music, Pandora, and Spotify have a lot of competition from other streaming music services. If you like to listen to music on your iPad, there are plenty of other options you can choose!

One of the biggest competitors is online retailer Amazon.com, which offers two different streaming music services. (That's in addition to all the digital music it has for purchase and download, of course.) Amazon Prime Music offers two million on-demand tracks for members of the Amazon Prime service. (Amazon Prime offers free two-day shipping on qualifying purchases; a one-year subscription costs $119 and includes free access to the Prime Music service.) If you want more selection, pay $7.99 per month (for Prime members) or $9.99 per month (for non-Prime members) for Amazon Music Unlimited, which offers 60 million on-demand tracks. Get the Amazon Music app for free from the App Store.

Tidal is a newer streaming music service that offers some exclusives from major artists and, with its higher level, better sound quality than other services. The site offers more than 60 million tracks (and 250,000 music videos); it is owned by a group of famous musicians, including Beyoncé, Kanye West, Jay-Z, and Rhianna. The Premium plan, with standard sound quality, runs $9.99 per month; the HiFi plan, with superior lossless sound quality, runs $19.99 per month. (Lossless audio is much higher quality than the typically "lossy" models embraced by most streaming services and the MP3 digital file format.)

Then there's YouTube Music. It's kind of like regular YouTube supplemented with millions of music tracks and music videos. You pay $9.99 per month for the service.

Speaking of YouTube, lots of people use the regular YouTube video service to listen to their favorite music—and it's all free. Most major artists and record labels have official music videos on YouTube, and many artists and individuals have created and uploaded so-called "lyric videos" (music with lyrics superimposed onscreen) for other songs. (Learn more about YouTube in Chapter 17, "Watching TV Shows, Movies, and Other Videos.")

Buying and Playing Music from the iTunes Store

While streaming music may be all the rage, many of us still prefer to own the music we like, so we can take and listen to it anywhere—even if there's no Internet connection handy. The way to do this on your iPad is to purchase digital music from Apple's iTunes Store and then download your purchases to your iPad for playback.

Purchase Music from the iTunes Store

Apple offers tens of millions of individual tracks for purchase in the iTunes Store. Most tracks are priced from $0.99 to $1.29.

(1) From the iPad Home screen, tap the iTunes Store icon to open the iTunes Store.

2 Tap the Music tab to view music available for purchase.

3 Tap Genres to browse music by genre.

4 Tap to select the genre you want. *Or…*

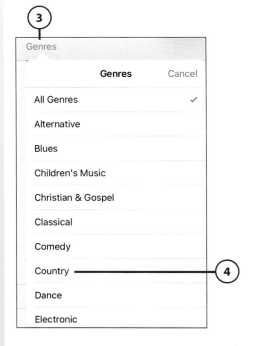

5 Tap within the Search box and enter the name of a song, album, or artist.

6 iTunes displays a list of matching artists or music. Tap the one you want.

7 Tap a song to purchase that track.

8 Tap an album to purchase the album or view its individual tracks.

9 Follow the instructions to complete the transaction.

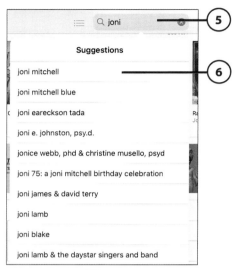

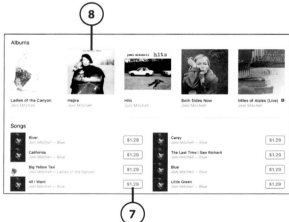

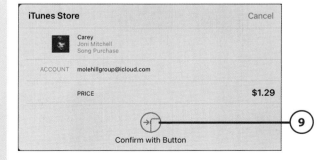

Play Your Tunes

When you purchase a track from the iTunes Store, that track is automatically downloaded to and stored on your iPad. You play this music with the iPad's Music app.

1. From the iPad Home screen, tap the Music icon to open the Music app.

2. In the Library section, tap to select how you want to view your music—Recently Added, Artists, Albums, Songs, Genres, Compilations, Composers, or Downloaded.

3. If you select Artists or Albums, tap through to view individual tracks.

4. Tap Play to play all the tracks in order.

5. Tap Shuffle to play all the tracks in random order.

6. Tap a track to play that track.

7. Playback controls appear at the bottom of the screen. Tap Pause to pause playback; tap Play to resume playback.

8. Tap Next to skip to the next track.

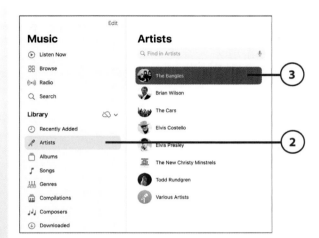

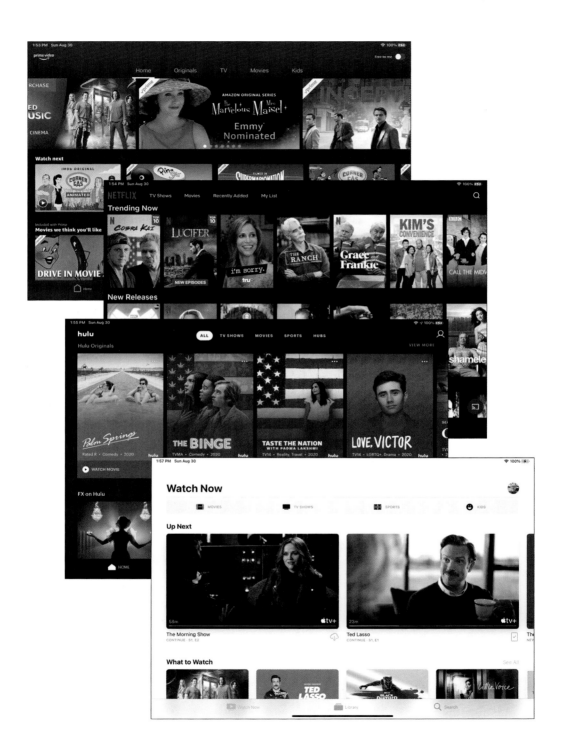

In this chapter, you learn how to use your iPad to watch movies, TV shows, and other videos.

→ Watching Paid Streaming Video Services
→ Watching Other Streaming Video Services
→ Using the Apple TV App
→ Watching YouTube

17

Watching TV Shows, Movies, and Other Videos

Your iPad is a great device for watching your favorite movies, TV shows, and other videos wherever you happen to be. Whether you're relaxing on your couch, sitting in a coffeehouse, or traveling cross country, just pull out your iPad and tap what you want to watch—on Amazon Prime Video, Apple TV+, Disney+, HBO Max, Hulu, Netflix, and other video services.

Watching Paid Streaming Video Services

Most of the video you watch on your iPad is *streaming video*. This is video that streams in real time over the Internet from a streaming video service. Of course, you also can purchase and download movies and TV shows to your iPad for future viewing, but most people these days watch their video entertainment via streaming video. There's nothing to download; just press "play" and you're ready to watch.

There are dozens of streaming video services today, but the biggest are Amazon Prime Video, Apple TV+, CBS All Access, Disney+, HBO Max, Hulu, and Netflix. All of these services are paid services—you need to subscribe and pay a monthly fee to these services to watch the content they offer.

Watch Amazon Prime Video

You probably know Amazon as a popular online retailer, but it's also a major player in streaming video, with lots of original content as well as popular movies and TV shows.

You watch Amazon Video from the Amazon Video app, which you can download for free from Apple's App Store. The Amazon Prime Video service is free if you have an Amazon Prime membership, or you can subscribe to just the video service for $8.99 per month. Of course, you also can purchase or rent individual programs and movies for downloading from Amazon's online store.

Amazon Prime

Amazon Prime Video is a part of Amazon Prime, the service that gives you free shipping on most Amazon orders. Prime Video is free with a Prime membership, which costs you $119 per year.

The first time you launch the Prime Video app, you're prompted to sign in with your Amazon account. Do so if you have one; if you don't yet have an Amazon account, you need to go to Amazon's website (www.amazon.com) and create one.

1. From the iPad's Home screen, tap the Prime Video icon to launch the Amazon Prime app.

2. Tap the Home tab to view Amazon Video's Home screen.

3. Tap the Originals tab to view Amazon original programming.

4. Tap the TV tab to view only TV shows.

5. Tap the Movies tab to view only movies.

6 Tap the Kids tab to view children's programming.

7 Tap Search to search for specific shows or movies.

8 Tap to select the item you want to watch.

9 If you selected a TV show, tap to select a season.

10 Tap an episode to expand the listing and view more details.

11 Tap the Play icon for the episode you want to watch.

12 If you selected a movie, tap Play Movie to watch now.

Widescreen Video

Because movies and TV shows are shot wider than they are tall (especially with true widescreen programming), the best way to watch movies on your iPad is to hold your device horizontally, in landscape mode.

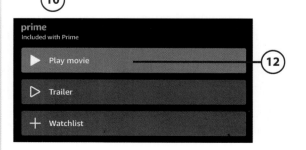

Watch Apple TV+

Apple TV+ (not to be confused with the Apple TV app or Apple TV streaming media player) is Apple's new streaming video service. It offers a selection of original series as well as some theatrical movies.

You subscribe to and watch Apple TV+ from within the Apple TV app. It costs $4.99 per month.

(1) Tap the Apple TV icon to launch the Apple TV app.

(2) Scroll to the Channels section and tap Apple TV+.

(3) Tap to select the item you want to watch.

(4) If you selected a series, scroll down to select the episode you want to watch.

(5) If you selected a movie, tap Play to begin playback.

Apple TV Streaming Media Player

Apple TV is also the name of Apple's streaming media device, a box that you connect to your TV to watch streaming video over the Internet. Learn more at www.apple.com/apple-tv-4k/.

Watch CBS All Access

CBS All Access offers content from the CBS television network (including CBS news and sports), as well as BET, Comedy Central, MTV, Nickelodeon, and the Smithsonian Channel. It also offers as movies from Paramount Pictures. (All channels are owned by the CBSViacom conglomerate.)

CBS All Access is also the home for a variety of original programming, including *The Good Fight, Star Trek: Discovery*, *Star Trek: Picard,* the upcoming *Star Trek: Strange New Worlds,* and Jordan Peele's reboot of *Twilight Zone.* You can also find all *Star Trek* series and movies here.

CBS offers two All Access plans. The Limited Commercials plan runs $5.99 per month or $59.99 per year. If you'd rather excise advertisements entirely, go with the No Commercials plan for $9.99 per month or $99.99 per year.

(1) Tap the CBS icon to launch the CBS All Access app.

(2) Tap a tile to view programming from that CBSViacom network—CBS, BET, Comedy Central, MTV, Nickelodeon, or Smithsonian Channel.

(3) Tap Search to search for specific programming.

(4) Tap Live TV to watch live television from your local CBS affiliate, CBS Sports HQ, the CBSN news channel, and ET Live.

(5) Tap Browse to browse through programming by type.

(6) Tap Shows to view television series.

(7) Tap Movies to view movies.

(8) Tap to select the item you want to watch.

(9) If you selected a television series, tap to select the desired season.

(10) Tap the thumbnail to play a specific episode.

(11) If you selected a movie, tap the Play button to start watching.

Paramount+

CBSViacom has announced that they intend to change the name of CBS All Access to Paramount+ to reflect the broad variety of non-CBS content available on the service. The name change should take place early in 2021.

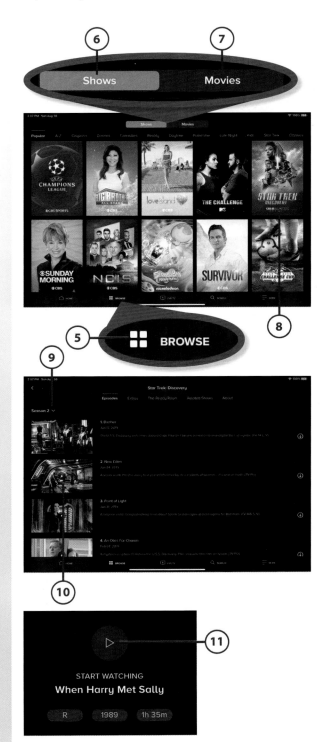

Watch Disney+

Disney+ is another new paid streaming service, featuring content from properties owned by the Walt Disney media conglomerate—specifically movies and television series from Disney, Marvel, National Geographic, Pixar, and 20th Century Fox. This includes all the Marvel superhero movies, programming from the Disney Channel and Disney Jr., classic and current Disney movies, and all the *Star Wars* movies and series.

That's a lot of programming, which makes it a favorite service for many, especially if you have younger viewers in your household. A subscription costs just $6.99 per month or $69.99 per year.

1 Tap the Disney+ icon to launch the Disney+ app.

2 If you've created multiple accounts for different members of your family, tap the icon for a specific user.

3 Tap the icon for the property you want to watch—Disney, Pixar, Marvel, Star Wars, or National Geographic. *Or…*

4 Tap Search to search for specific content.

5 Tap to select the item you want to watch.

6 If you selected a television series, tap to select a season.

7 Tap a thumbnail to play that episode.

8 If you selected a movie, tap Play to begin playback.

My TV for Seniors

Learn more about streaming video on your iPad and on your TV in my companion AARP book, *My TV for Seniors*, available wherever books are sold.

Watch HBO Max

HBO Max offers content from all the properties owned by parent company AT&T—Cartoon Network, CNN, The CW, DC Comics, HBO, New Line Cinema, Turner Classic Movies (TCM), and the Warner Bros. film studio. Despite the HBO name, HBO Max is designed to be a full-service streaming service, much like Netflix, with a mix of existing and original programming.

A monthly subscription to HBO Max runs $14.99 USD, although you may get a discount if you also subscribe to HBO through your cable provider.

1 Tap the HBO Max icon to launch the HBO Max app.

2. Tap the Search icon to search for specific content. *Or…*

3. Scroll down to the HBO Max hubs section and select a hub—HBO, DC, Sesame Workshop, TCM, Studio Ghibli, Cartoon Network Collection, adult swim connection, Crunchyroll Collection, or Looney Tunes.

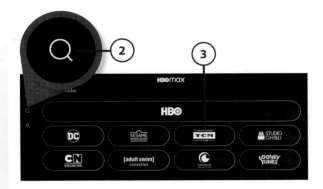

4. Tap to select the item you want to watch.

5. If you selected a series, tap to select a season.

6. Tap a thumbnail to watch a specific episode.

7. If you selected a movie, tap the Play button.

Watch Hulu

Hulu is a streaming video service known for offering recent episodes of network and cable programming. You can watch just about any TV program within a few days of its original airing.

Hulu also offers a variety of original programming, including *Castle Rock*, *Casual*, *Devs*, *Future Man*, *The Handmaid's Tale*, *High Fidelity*, *Letterkenny*, *The Looming Tower*, *Runaways*, and *Shrill*. Hulu is also a good place to find a variety of vintage television programs.

Hulu offers two plans, with and (sort of) without commercials. The basic Hulu plan runs $5.99 per month and inserts commercials into the programs you watch. If you want to minimize the number of commercials you see, sign up for the $11.99 No Ads plan—but know that you'll still see commercials on some programs, but fewer of them.

1. Tap the Hulu icon to launch the Hulu app.

2. Tap the Home icon to view Hulu's Home screen.

3. Tap All to view all available programming.

4. Tap TV Shows to view only television programming.

5. Tap Movies to view only movies.

6. Tap Sports to view only sports programming.

7. Tap Hubs to view programming by network or service.

8. Tap Search to search for specific content.

9. Tap to select the item you want to watch.

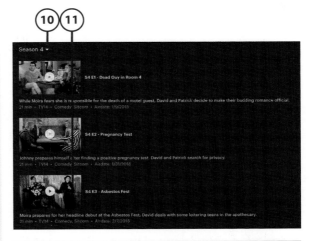

10. If you selected a TV program, tap to select a season.

11. Tap a thumbnail to watch that episode.

12. If you selected a movie, tap Watch Movie to begin watching.

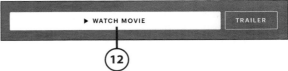

Watch Netflix

Netflix is the most popular streaming video service today. It offers a variety of movies, past and current TV shows, and original programming.

Netflix offers three subscription plans: Basic ($8.99 per month), which doesn't offer high-definition (HD) viewing; Standard ($12.99 per month), which *does* offer HD, plus the ability to view on two different devices at the same time; and Premium ($15.99 per month), which lets you watch on four different screens simultaneously. For most users, the Standard plan is the one to get.

1. Tap the Netflix icon to launch the Netflix app.

2. Netflix lets you create multiple viewer profiles. If you've created more than one profile, they're displayed on the Who's Watching? screen. Tap the profile you want to use.

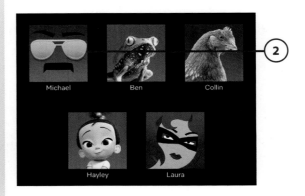

(3) On the Netflix home screen, scroll down to see various sections: New Releases, Trending Now, Watch It Again, and so forth.

(4) Tap TV Shows to view only television shows.

(5) Tap Movies to view only movies.

(6) Tap Recently Added to view programming just added to Netflix.

(7) Tap My List to view items you've added to your favorites list.

(8) Tap the Search (magnifying glass) icon to display the Search page and search for specific programming.

(9) Tap the program or movie you want to watch.

(10) If you selected a television show, you see a detail panel for that show. Tap the Seasons down arrow to select all episodes from a given season.

(11) Tap a thumbnail to start watching that episode.

(12) If you selected a movie, you see the detail panel with information about that movie. Tap the Play icon to begin watching.

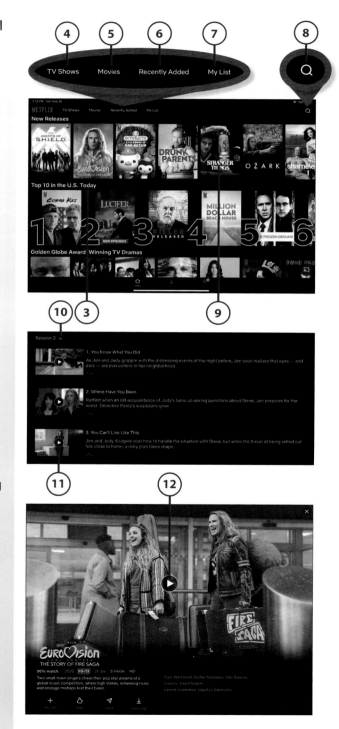

Watching Other Streaming Video Services

The streaming video services just discussed may be the most popular today, but they're far from the only ones. There are literally dozens of other streaming video services available for viewing on your iPad.

Discover Other Streaming Video Services

Table 17.1 details some of the more popular services; you can access all of these with apps available for downloading from Apple's App Store.

Table 17.1   Other Streaming Video Services

Service	Description	Monthly Subscription
Acorn TV	Programming from the UK, Canada, New Zealand, and around the world	$5.99
BET+	Movies and TV shows from Black creators for Black audiences	$9.99
BritBox	UK programming from the BBC and ITV	$6.99
BroadwayHD	Broadway shows and concerts	$8.99
Crackle	Older movies and TV shows	Free
The Criterion Channel	Classic films from the Criterion Collection	$10.99
ESPN+	Sports programming from ESPN's family of networks	$5.99
Fawesome	Theatrical and TV movies	Free
FilmRise	Movies, '80s and '90s TV shows	Free
Peacock	Programming from the NBC network (including NBC News and NBC Sports), DreamWorks Animation, and Universal Pictures	Free, Premium ($4.99), and Ad-Free Premium ($9.99)
Plex	Older movies and TV series	Free

Service	Description	Monthly Subscription
Popcornflix	Older movies, reality TV shows, children's programming	Free
Pluto TV	TV shows, movies, and children's programming, plus national news and sports	Free
Stirr TV	Older movies and TV shows, plus local and national news	Free
Tubi	Older movies and TV shows, reality TV shows	Free
Xumo	Reality and documentary movies and TV shows, plus national news	Free

Watch Live TV Online

Beyond traditional streaming services are *live* streaming services that let you watch live TV, from both local channels and cable/satellite channels, on your iPad. This type of service displays a program guide, like the kind from a cable or satellite provider, with available channels listed by time of day.

The program guide for Hulu + Live TV

None of these services are free, however; all have a monthly subscription fee. Table 17.2 details the major live streaming services with iPad apps.

Table 17.2   Live Streaming Video Services

Service	Monthly Subscription
AT&T TV	$59.99+
fuboTV	$64.99+
Hulu + Live TV (uses Hulu app)	$54.99+
Sling TV	$30.00+
YouTube TV	$64.99+

Using the Apple TV App

Apple makes streaming video on your iPad easy with the Apple TV app. The Apple TV app tries to be a central hub for content from multiple paid streaming video services. You select the content you want in the Apple TV app, then your iPad launches the app for that service for your viewing pleasure.

Find Something to Watch

The Apple TV app presents content from a variety of paid streaming services as well as the iTunes Store in a unified interface. When you select a program or movie to watch, the app launches the app for the service where that content resides. You then watch the selected content on the other app.

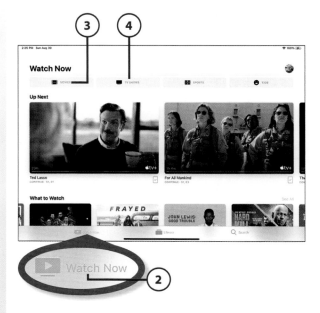

1. Tap the Apple TV icon to launch the Apple TV app.

2. Tap the Watch Now tab.

3. Tap Movies to view available movies.

4. Tap TV Shows to view available television programs.

(5) Tap Sports to view available sports content.

(6) Tap Kids to view available children's programming.

(7) Tap Search to search for programs.

(8) Scroll down to view suggested programming.

(9) Tap to select an item.

(10) If the program is from a streaming service you subscribe to, tap the Play button to watch.

(11) If the program is from a streaming service you don't yet subscribe to, you're prompted to either get the app (for free services) or subscribe.

(12) If the program is available for sale or rental from the iTunes Store, tap the Buy or Rent button.

Free Trials

Many subscription services offer free trials so you can check out what's available before committing to a paid subscription.

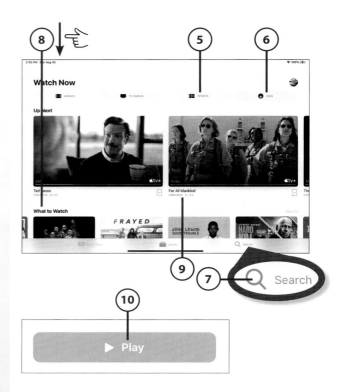

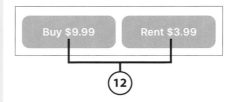

Subscribe to a Channel

You can use the Apple TV app to manage subscriptions to other paid streaming video services—what are called *channels* within the app. You can choose to subscribe to a channel within the Apple TV app or cancel any subscriptions you may have.

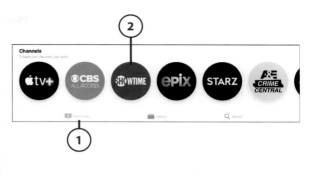

(**1**) From within the Apple TV app, tap to select the Watch Now tab.

(**2**) Scroll to the Channels section and tap to select the channel to which you'd like to subscribe.

(**3**) Tap the Try Free button to sign up for a free trial. (Most free trials are for a week or so.) You will be charged the full subscription price at the end of the trial.

Manage Your Subscriptions

When you subscribe to a channel through the Apple TV app, your subscription is billed to your Apple account. You can unsubscribe from channels at any time through the Apple TV app.

(**1**) From within the Apple TV app, tap your profile picture in the top-right corner to display the Account pane.

(**2**) Tap Manage Subscriptions to display the Subscriptions pane and see a list of all your subscriptions.

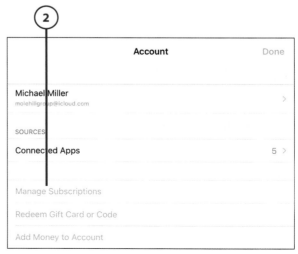

(3) Tap the subscription you'd like to manage.

(4) Tap Cancel Subscription.

(5) When prompted to confirm the cancellation, tap Confirm.

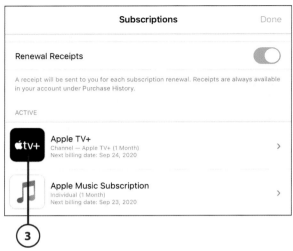

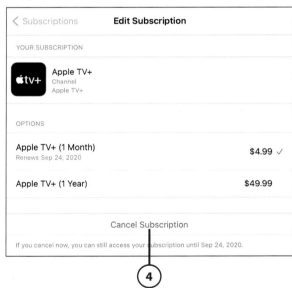

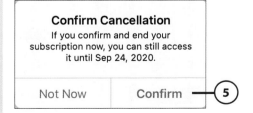

Watching YouTube

Then there's YouTube. One of the more popular sites on the Web, YouTube is a video-sharing community where regular people, like you and me, can upload and share their home movies and other videos. There's also a lot of professional content on YouTube, including tons of music videos and educational videos.

Unlike most other streaming video services, YouTube is completely free. (It's ad supported.) If you want to take advantage of all of YouTube's features, including your personal playlists and favorites, you need to sign in with your Google account.

Find a Video

You watch YouTube from the YouTube app, available (for free) from Apple's App Store.

1 Tap the YouTube icon to launch the YouTube app.

2 Tap the Home tab to view recommended videos.

3 Tap the Explore tab to browse videos by category (Trending, Music, News, Learning, and so forth).

4 Tap the Subscriptions tab to view videos from channels to which you've subscribed.

5 Tap the Search icon to search for specific videos.

Watch a Video

Watching a YouTube video is much like playing a movie or TV show from a traditional video streaming service.

1. Tap a video's thumbnail image or title to play the video.

2. Tap the thumbs up icon to like the video.

3. Tap the thumbs down icon if you don't like the video.

4. Tap the Share icon to share the video via text, email, or social media.

5. Tap Subscribe to subscribe to this YouTuber's channel and get notified of new videos from this person.

6. Tap the screen to display the playback controls and then tap Pause to pause playback. Tap Play to resume playback.

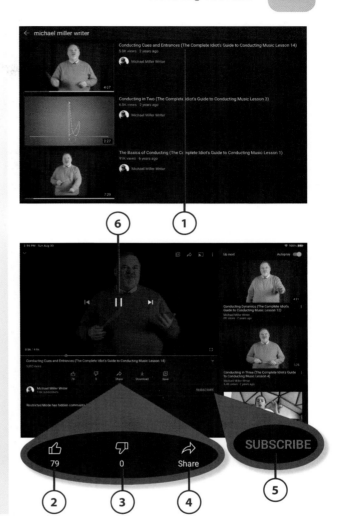

>>>*Go Further*
BUYING AND RENTING VIDEOS

If you can't find a TV show or movie you like on a streaming service, you may be able to buy or rent it from Apple's iTunes Store. You access the iTunes Store from the iTunes Store app.

The items you rent or purchase can be stored in your iCloud account or downloaded to your iPad so you can watch when you don't have an Internet connection, like on a plane or bus ride. Use the iTunes Store app to make your purchases, and use the Apple TV app to watch what you've bought or rented.

In this chapter, you learn how to use Apple HomeKit and the Home app on your iPad to control smart devices in your home.

→ Understanding Apple HomeKit
→ Adding Devices and Rooms
→ Controlling Your Smart Devices

18

Controlling Your Smart Home

One of the big trends today is setting up so-called "smart" devices to control various functions in your home. We're talking smart lighting that turns on and off according to a preset schedule, smart thermostats that learn your behavior and adjust the temperature accordingly, and smart locks and doorbells that let you see who's at the door and unlock the door if you want to, all from a corresponding app on your iPad or smartphone.

Most smart devices come with their own mobile apps. This is fine—unless you have multiple smart devices, in which case you have to deal with multiple apps. You can get around this multiple-app mayhem by using Apple's Home app on your iPad. The Home app works with Apple's HomeKit protocol to control multiple smart devices from multiple manufacturers.

If you'd like more information about using smart home devices and apps, check out Que's companion AARP book, *My Smart Home for Seniors*, available wherever books are sold.

Understanding Apple HomeKit

Apple HomeKit is a system for controlling smart devices from a variety of suppliers with a single app—or with Apple's Siri voice-controlled personal assistant. If a smart device is compatible with HomeKit (not all are), it's a simple matter to link the device to the Home app on your iPad (or iPhone) and control it along with other devices with a tap of your finger.

It's actually kind of a big deal to have centralized control of multiple smart devices with a single app. For one thing, using one app is a lot more convenient than using multiple apps. Beyond convenience, however, HomeKit also lets you centralize control of multiple devices in a single room, as well as create *scenes* that control multiple connected devices with a single tap. HomeKit makes possible a lot of combined automation that makes your smart devices work smarter together.

How HomeKit Works

HomeKit isn't a smart home system per se; it doesn't operate from a central hub, as do a lot of similar systems. Instead, HomeKit is more like the standalone smart controllers offered by Amazon (Echo) and Google (Home), but it works through a smartphone/tablet app instead of a piece of hardware (although you can control HomeKit devices, after they've been set up, with Apple's HomePod smart speaker).

So how does HomeKit work? It all revolves around Apple's Home app, which comes installed on your new iPad.

If a smart device is compatible with the HomeKit standard, it's typically discovered by Apple's Home app when you first power on the device. (If it's not automatically discovered, you can add it manually to the app.) After a device is linked to the Home app, you specify what room it's in. You can then control the device from the Home app or via any scenes or automations you create. (Your iPad must be connected to your home Wi-Fi network to do this; Home can't control your smart devices remotely.)

What's Compatible?

To work with HomeKit, smart devices need to be certified by Apple. Many are, but some aren't. To determine what's compatible, see the list at www.apple.com/ios/home/accessories/.

Adding Devices and Rooms

Before you can use the Home app to control devices from your iPad, you have to add those devices to the app. You can also assign devices to specific rooms in your home.

Apple calls smart devices, whether light bulbs or thermostats, "accessories." If you have existing smart home devices, you're prompted to set them up when you launch the Home app for the first time. When you see the Welcome screen, tap Continue and then follow the onscreen directions. You're prompted to add your devices and create rooms. You can then add other devices and rooms manually at any time.

Add a New Accessory

Compatible accessories are relatively easy to add to the Home app. Just make sure the accessory is powered on and connected to your home's Wi-Fi network—and that your iPad is connected to the same network.

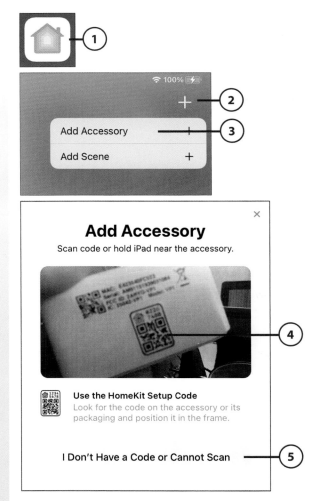

Add Accessory

Scan code or hold iPad near the accessory.

Use the HomeKit Setup Code
Look for the code on the accessory or its packaging and position it in the frame.

I Don't Have a Code or Cannot Scan

(1) Tap the Home icon to launch the Home app.

(2) From the Home screen, tap the + sign.

(3) Tap Add Accessory.

(4) Look for the HomeKit code on the new device or in the device's documentation. Position the camera of your iPad to capture this code.

(5) If for some reason step 3 doesn't work, tap I Don't Have a Code or Cannot Scan to enter the code with your tablet's onscreen keyboard.

(6) Tap Add to Home.

(7) Select which room you want to add this accessory to.

(8) Tap Continue.

(9) Accept the name of the accessory or give it a new, more descriptive one.

(10) Tap Continue.

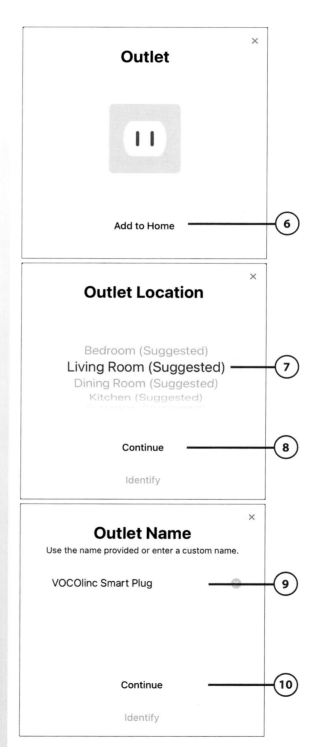

×

Outlet

Add to Home ——— (6)

×

Outlet Location

Bedroom (Suggested)
Living Room (Suggested) ——— (7)
Dining Room (Suggested)
Kitchen (Suggested)

Continue ——— (8)

Identify

×

Outlet Name

Use the name provided or enter a custom name.

VOCOlinc Smart Plug ——— (9)

Continue ——— (10)

Identify

11. Choose how you want the accessory to display within the Home app.

12. Tap Continue.

13. Tap Done to complete the installation.

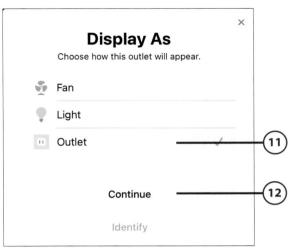

Display As

Choose how this outlet will appear.

Fan

Light

Outlet ✓ — 11

Continue — 12

Identify

Outlet Added to My Home

Done — 13

View in Home

Create a Scene

With the Home app, you can control multiple accessories at the same time by creating what Apple calls *scenes*.

For example, you can create a scene called "Good Morning" that turns on the smart lights in your bedroom and turns up your smart thermostat, or one called "Arrive Home" that turns on your living room lights and unlocks the front door. You activate a scene with a single tap from the home or room screen or via Siri command.

1. On the Home screen, tap the +.

2. Tap Add Scene.

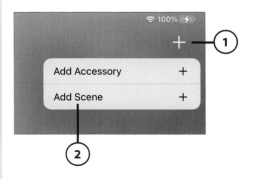

🔋 100% 🔋

\+ — 1

Add Accessory   +

Add Scene   +

— 2

3 Choose one of the suggested scenes. *Or…*

4 Tap Custom to create a new scene.

5 Enter a name for this scene.

6 Tap Add Accessories.

7 Tap those accessories you want to add.

8 Tap Done.

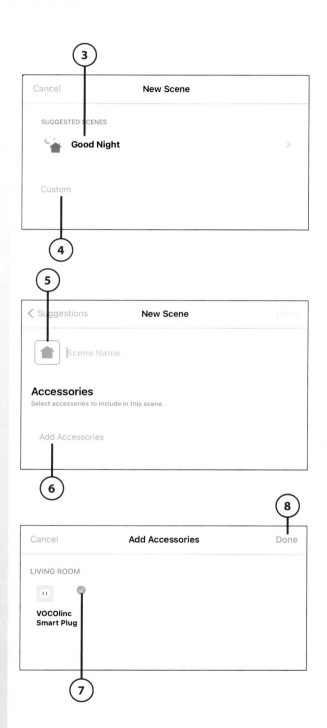

(9) Press and hold an accessory to select what that accessory does during the scene.

(10) Tap Add Accessories to add other accessories to this scene.

(11) Tap Test This Scene to see if the scene works as you want.

(12) Make sure the Include in Favorites switch is "on" to display this scene on the home screen.

(13) Tap Done to add this scene.

Controlling Your Smart Devices

Once you've installed a device/accessory in the Home app, you can use your iPad to control that device in various ways.

Control a Device or Scene from Your iPad

Turning a device on or off, or activating a scene, is as easy as tapping your tablet's screen. For many devices, additional controls are also available.

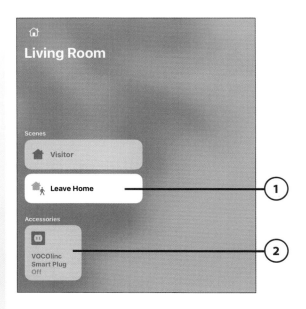

(1) On either the Home or Rooms screen, tap an accessory or scene to turn it on or off. The tile for the accessory or scene dims when the device is turned off; the tile has a white background when the device is turned on.

(2) Press and hold the tile for an accessory to access additional controls for that item.

Control Devices and Scenes with Siri

You're probably used to asking your iPad's Siri personal assistant to do all manner of things, from looking up information to reading the latest news and weather reports. Well, you can also use Siri within the Home app to control your connected smart devices using voice commands.

To use Siri to control any devices and scenes you've added to the Home app, all you have to do is say "Siri," then speak the desired command. It's that easy.

Obviously, which commands you speak depend on which devices you have installed and what you want them to do. For example, if you want to turn off the lights in your bedroom, say this:

Siri, turn off the lights in the bedroom.

Pretty simple. If you want to reduce the brightness for a given light or room, say this:

Siri, set brightness for the living room lights to 50 percent.

To set your smart thermostat, say this:

Siri, set the temperature to 70 degrees.

You can even ask questions of Siri regarding your smart devices, like this one:

Siri, did I lock the front door?

If you have scenes created (refer to the "Create a Scene" task earlier in this chapter), you can turn on a scene by saying

Siri, set my relaxing scene.

It's all very common sense. All you have to do is speak to Siri via your iPad, and she works within the Home app to do the rest.

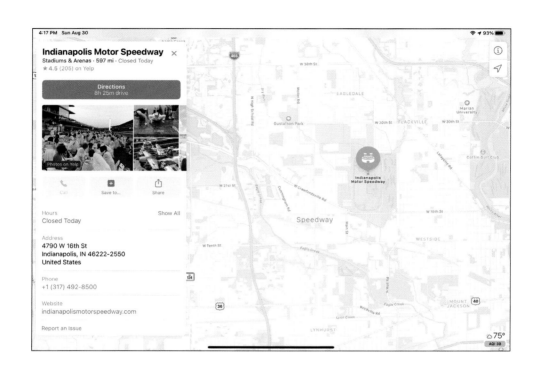

19

In this chapter, you learn how to use your iPad when you're traveling, whether near or far.

→ Traveling with the Maps App
→ Exploring Other Travel-Related Apps

Traveling with Maps and Other Travel Apps

Your iPad is a handy device both at home and when you're traveling. You can use the Maps app to generate maps and driving directions, and there are all sorts of other fun and useful travel-related apps you can use when you're away from home.

Traveling with the Maps App

Apple's Maps app can show you how to get just about anywhere from wherever you are. It displays maps of any location, as well as driving directions for how to get there.

Other Driving Apps

Apple's Maps app isn't the only driving/map app available. There's also the popular Google Maps app; it offers a few more features and is sometimes more accurate than Apple's Maps app. Waze Navigation & Live Traffic is also popular because it offers crowd-sourced live traffic info, including local tie-ups, to help you find the fastest route to where you're going.

Display a Map

To use the Maps app, enter any location—street address, intersection, major landmark—and it will generate a map of that location.

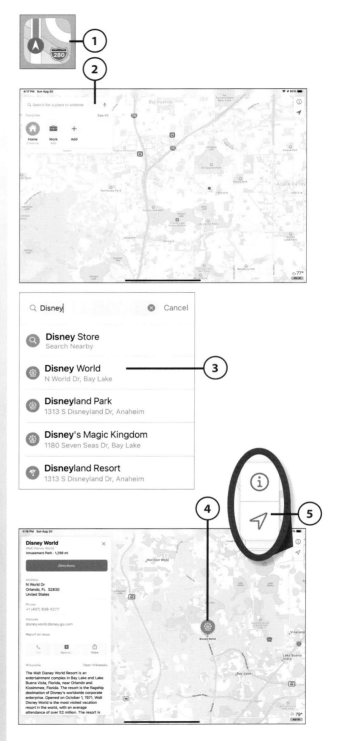

1. Tap the Maps icon to open the Maps app. You see a map of your current location.

2. Tap within the Search panel and enter the new location you want mapped. You can enter a street address (accompanied by city and state, if you like), intersection (10th and Main, for example), landmark (such as Brooklyn Bridge or Wrigley Field), or business.

3. As you type, the Maps app displays suggested matches. Tap the one you want to map, or continue typing and then tap Search on the onscreen keyboard.

4. The app maps the location you entered, which is pinpointed in the middle of the map. (If you mapped a public location, there's also an information panel on the left.) Use your fingers to drag the map left, right, up, or down, or expand/pinch your fingers to zoom in to and out of the map.

5. Tap the Home (arrow) icon to return to (and map) your current location.

Generate Driving Directions

One of the most useful aspects of the Maps app is to generate driving directions from where you are to a given location.

(1) From within the Maps app, tap within the Search panel and enter the address or description of the destination location.

(2) The location is mapped, and a descriptive panel appears on the left side of the screen. Tap the Directions button.

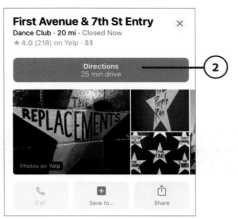

(3) Suggested routes are displayed from your current location. (To start your journey from a different location, tap My Location and enter a different address.) Tap the route you want to take; this route is mapped onscreen in dark blue. (Alternate routes are in light blue.)

(4) Tap the Go button for your selected route to display turn-by-turn instructions.

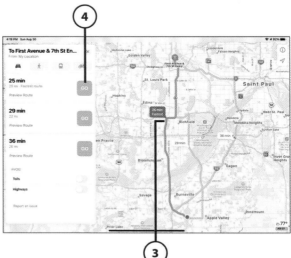

Walking, Biking, and Mass Transit

Maps can generate directions if you're walking, biking, or taking mass transit. Just tap the Walk, Bike, or Transit icons in the directions panel to change your mode of transport.

5 Swipe right to left in the instructions pane to advance to the next step. Swipe from left to right to return to the previous step.

6 Tap End to cancel the directions and return to the Where Do You Want to Go? panel.

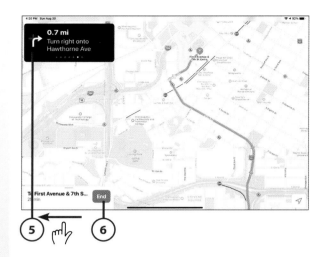

Maps in Your Car

If you have an iPad with cellular connectivity, you can use the Maps app in your car to generate turn-by-turn directions while you drive. Because most iPads have only Wi-Fi connectivity, this may not be an option for you. Instead, use the iPad to preview your route while you're still inside and connected to Wi-Fi.

Exploring Other Travel-Related Apps

The Maps app isn't the only iPad app useful to travelers. This section takes a look at other travel-related apps you might want to consider.

Discover Travel Apps

Here are just a few of the many travel-related iPad apps in Apple's App Store. You can search for specific apps by name, or search for *travel* to see all available apps.

1 Find and book hotel rooms, airline flights, rental cars, and more with the Booking.com, Expedia, Hotels.com, HotelTonight, Kayak, Orbitz, Priceline, Skyscanner, Travelocity, Trivago, and similar apps.

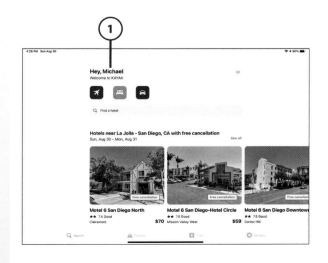

2 Use the TripAdvisor app to learn more about what's available at a given destination—and book your travel plans, too.

3 Use a hotel chain's app to book lodging directly.

4 Use airline apps to book flights directly.

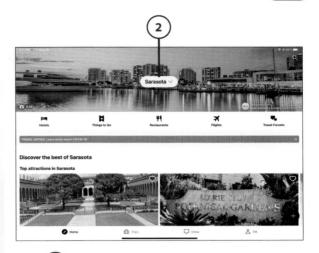

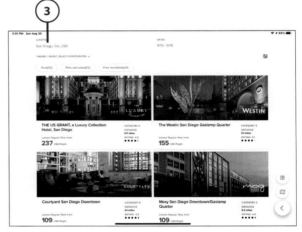

5 Use the Yelp app to find and read reviews for local restaurants.

6 Use the OpenTable app to make a restaurant reservation online.

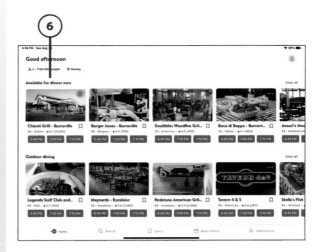

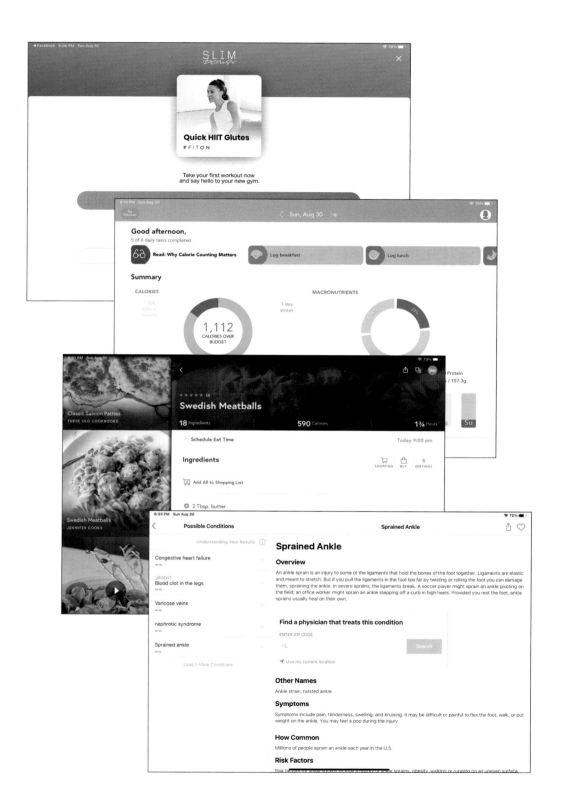

In this chapter, you discover a variety of health and fitness apps for your iPad.

→ Exploring Health and Fitness Apps
→ Exploring Food and Nutrition Apps
→ Exploring Medical Apps

20

Keeping Fit and Healthy

Many people use their iPads to help them manage their daily health and fitness needs. You can find workout and yoga apps, food and nutrition apps, even apps to help you find important medical information. These apps add more value to your iPad and can help you keep healthy!

Exploring Health and Fitness Apps

When it comes to staying fit, nothing beats the combination of a healthy diet and a regular exercise routine. I cover diet-related apps in the next section, but for now the focus is on the fitness end of things.

Discover Exercise and Yoga Apps

We all know how important it is to get enough exercise. Here are some of the more popular apps you can use to track your daily exercise and guide your workouts, available from the App Store.

(1) Daily Workouts Fitness Trainer (free) provides ten different 5- to 10-minute targeted workouts for both men and women.

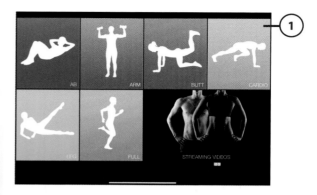

(2) FitOn: Workouts & Workout Plans (free) helps you get fit and lose weight. It includes a variety of different workouts, including cardio, HIIT, strength, yoga, Pilates, and dance.

(3) Daily Yoga: Workout & Fitness (free with in-app purchases) gets you started with a variety of entry-level exercises.

Finding Health and Fitness Apps

You can find all the health and fitness apps discussed in this chapter—and more!—in Apple's App Store. Just launch the App Store app and search for a particular app by name or type.

It's Not All Good

In-App Purchases

Many purportedly "free" apps offer in-app purchases for additional features or content. Just because you get an app for free doesn't mean it won't try to charge you anything!

Exploring Food and Nutrition Apps

When it comes to staying healthy, exercise is just part of the equation. We also need to watch what we eat—which we can do with these dieting and food apps.

Discover Dieting Apps

Check out some of these more popular apps to keep track of what you eat—and eat healthier!

1. Lose It! (free) is a calorie-counting app. Set your weight-loss goals and then track your foods to lose weight.

2. MyFitnessPal's Calorie Counter & Diet Tracker (free) is an easy-to-use calorie counter with a huge database of more than 11 million different foods.

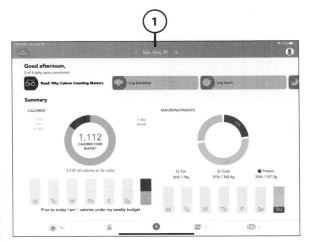

3 When you want to eat healthy, MyPlate Calorie Counter (free) tracks everything you eat with a database of more than 2 million food items.

Discover Cooking and Recipe Apps

Whether you're trying to eat healthy or just want to spice up your dinner routine, check out some of these popular cooking apps for all types of cuisine.

1 Use the Allrecipes Dinner Spinner (free) to access popular recipes from the 30 million members of the Allrecipes community—and watch more than 1,000 step-by-step cooking videos, too!

2 The Tasty app (free) offers more than 4,000 recipes along with an innovative step-by-step instruction mode and a search tool you can use to filter by ingredients, cuisine, and social occasion.

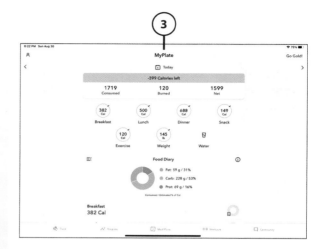

(**3**) Yummly Recipes and Shopping List (free) offers more than 2 million unique recipes and lets you create your own menu plans and shopping lists.

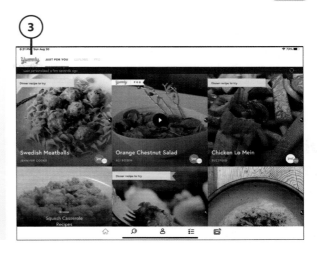

Exploring Medical Apps

There's a ton of medical services and information available on the Internet. The following apps let you access that information—and better track your own medical needs.

Discover Medical-Related Apps

The following apps let you access large online databases about medical conditions, symptoms, and medications. (You still should contact your own personal physician if you think there's something wrong, of course!)

(**1**) The Drugs.com Medication Guide app (free) lets you look up prescription drug information, identify any random pills you might have lying around the house, check drug interactions, and manage your prescription records. Other popular medication-related apps: Coverage Search (free) and Pocket Pharmacist (free).

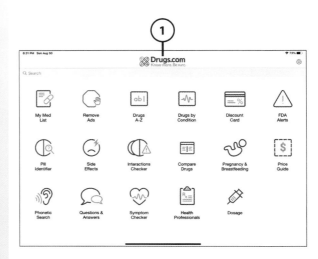

(2) Medical Dictionary by Farlex (free) is a medical dictionary app with more than 180,000 medical terms, 50,000 audio pronunciations, and 12,000 images.

(3) The WebMD app (free) offers a huge database of medical conditions, a symptom checker, medication reminders, drug information, first-aid essentials, and local health provider listings.

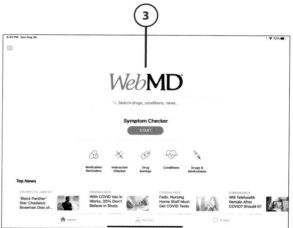

It's Not All Good

Don't Self-Diagnose

As tempting as it is to plug your symptoms into one of these medical-related apps and see what pops out, these apps should never replace the experience, knowledge, and advice you get from your physician. Although these apps can be informative, they're not always accurate or complete. It's always best to consult with your doctors—and use these apps to supplement, not replace, that information.

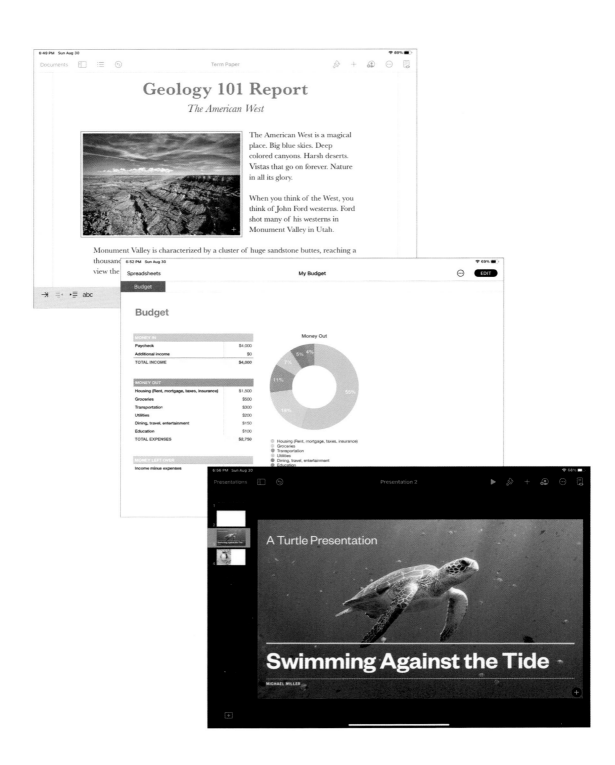

In this chapter, you learn how to use your iPad for business with key productivity apps.

→ Using Your iPad's Built-In Productivity Apps
→ Discovering More Productivity Apps

21

Getting Productive

It might not look like it, but your iPad is as powerful as many note-book computers (and especially powerful if you have an iPad Pro). That means you can use your iPad for just about any office task, from word processing to spreadsheets to presentations. All you need are the right apps—some of which come with your iPad and others available from the App store.

Using Your iPad's Built-In Productivity Apps

Your iPad comes with three apps that can help you be more produc-tive in the office and beyond. These apps are Pages for word process-ing, Numbers for spreadsheets, and Keynote for presentations. These free apps, which are comparable to and file-compatible with similar Microsoft Office apps, serve the day-to-day productivity needs of many users.

iPad Pro for Productivity

Although the standard iPad is the best solution for most home users, productivity apps work even better if you have an iPad Pro. The Pro's larger screen and more powerful processor, along with the Magic Keyboard or a similar third-party keyboard, make office work as easy and intuitive as it is on a traditional notebook computer.

Write with Pages

The Pages app is a word processor much like Microsoft Word, but it's optimized for use on the iPad. You can use Pages with your iPad's onscreen keyboard or, even better, use a wireless keyboard for faster and more accurate typing.

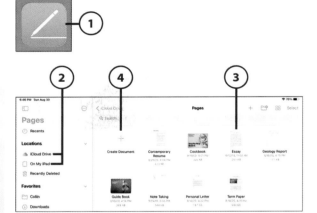

1 Tap the Pages icon to launch the Pages app.

2 In the Pages panel on the left, tap either iCloud Drive or On My iPad to view documents stored in the cloud or on your device.

3 Tap a tile to reopen an existing document. *Or…*

4 Tap Create Document (+) to create a new document.

5 Tap to select a template for your new document.

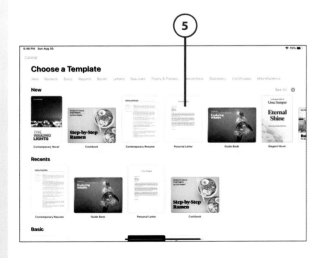

6 Most documents include place-holder text as defined by the template you selected. Tap any block of placeholder text to replace that text with your own text.

7 Tap the paintbrush icon in the toolbar to format the selected text.

8 Tap the Paragraph Style control to change the paragraph style.

9 Tap Font to change the font family.

10 Tap B to boldface text.

11 Tap I to italicize text.

12 Tap the – and + buttons to decrease or increase the size of the selected text.

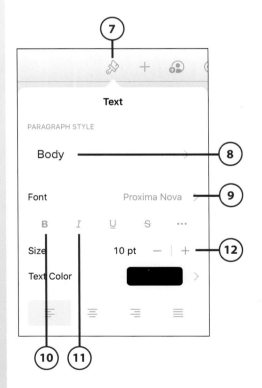

13 Tap the + icon to insert a table, chart, shape, photo, or video.

14 Tap the More (three-dot) icon to share a copy of this document, print this document, and more.

Stored in the Cloud

By default, all Pages, Numbers, and Keynote documents are stored in Apple's iCloud. This means you don't have to manually save your work; all changes are saved automatically as you make them. It also means that your documents won't fill up the limited storage on your iPad, and they will be available to any device connected to the Internet via your iCloud account.

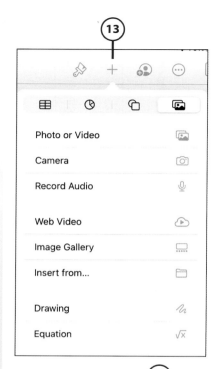

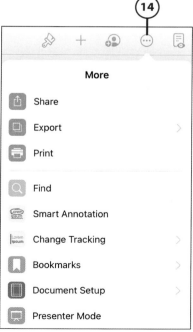

Work with Numbers

You use the Numbers app to... well, work with numbers. You can create spreadsheets in which you can enter and manipulate numerical data. It's great for creating budgets, expense reports, financial plans, and the like. You can perform all sorts of complex calculations and then display your results in attractive charts and graphs.

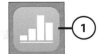

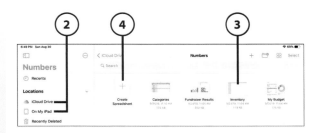

1. Tap the Numbers icon to launch the Numbers app.

2. In the Numbers panel on the left, tap iCloud Drive or On My iPad to view spreadsheets stored in the cloud or on your device.

3. Tap one of these tiles to reopen an existing spreadsheet. *Or...*

4. Tap Create Spreadsheet (+) to create a new spreadsheet.

5. Tap to select the template you want to use for your new spreadsheet.

6. Tap any cell to enter text or numbers. If a cell includes placeholder contents, tap to enter your own text or numbers. (To select more than one cell, use your finger to drag any of the selection corners to include additional cells.)

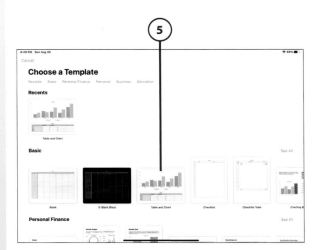

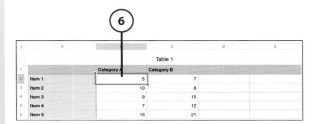

7 Tap the paintbrush icon to format the selected cell(s).

Rows, Columns, and Cells

A spreadsheet consists of multiple *rows* (arranged down the left edge of the screen) and *columns* (arranged along the top of the screen). The intersection of any given row and column is called a *cell*; all the contents of a spreadsheet are contained in individual cells.

8 Tap the Table tab to format the selected cells as a table.

9 Tap the Cell tab to format the contents of the selected cell(s). You can apply text formatting (bold, italic, underline, strikethrough); change text options (size, color, and font); apply cell fill color; add various types of borders; and opt whether to wrap long text in a cell.

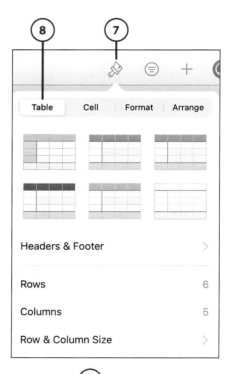

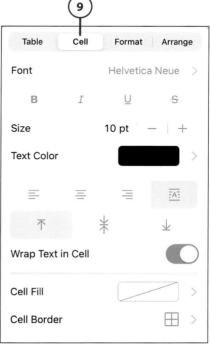

10 Tap the Format tab to format the numerical data in the selection. You can choose a general Number format, Currency format, Percentage, and so forth.

11 Tap the Arrange tab to move the selected element forward or backward on the page.

12 Tap the + icon to insert a table, chart, shape, or picture.

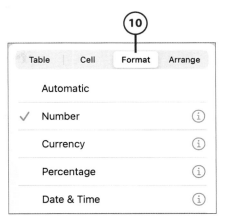

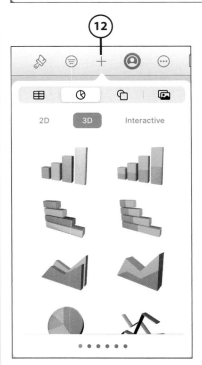

13 Tap a cell to manually enter a formula or select a function to perform a numerical calculation.

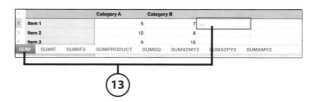

Formulas and Functions

You perform a calculation in a spreadsheet by entering a formula that may include the contents of specified cells. You can include built-in formulas, called functions, within your larger formulas. (Unfortunately, the whole concept of formulas and functions is much more involved than I can discuss in this chapter. For onscreen coaching tips, tap More and select Numbers Help.)

Present with Keynote

The Keynote app lets you create highly visual presentations that you can present on your iPad or by connecting to a larger monitor or projector.

1 Tap the Keynote icon to launch the Keynote app.

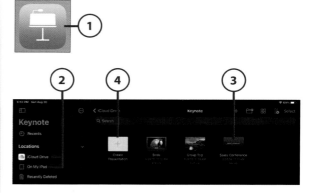

2 In the Keynote panel on the left, tap either iCloud Drive or On My iPad to view presentations stored in the cloud or on your device.

3 You see tiles for presentations you've previously created. Tap one of these tiles to reopen an existing presentation. Or...

4 Tap Create Presentation (+) to create a new presentation.

5 Tap to select the template you want to use for your new presentation.

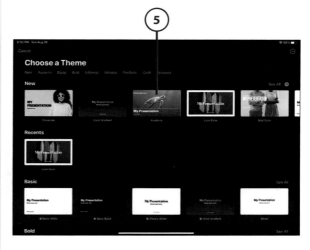

(6) The new presentation includes a preselected visual theme, along with a single opening slide. The slides in your presentation are displayed in the slide sorter on the left side of the screen. Tap any slide to view it in the larger editing area.

(7) Each new slide includes placeholder text and graphics. Double-tap any placeholder to replace its contents with your own.

(8) Select a block of text and then tap the paintbrush icon to format the text's Style (background and text colors), format selected Text (font family and size, bold and italic formatting, and text alignment), and Arrange elements on the screen.

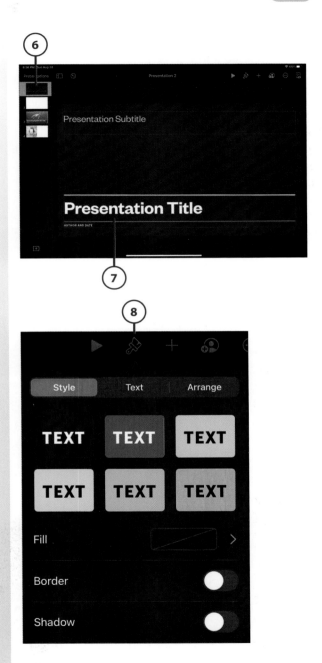

9 Tap the + icon to insert a table, chart, shape, or picture.

10 Tap the New Slide (+) icon in the lower-left corner to add a new slide to the presentation.

11 Tap to select the type of slide you want to add. (Repeat the preceding steps to add content to the slide and add additional slides to the presentation.)

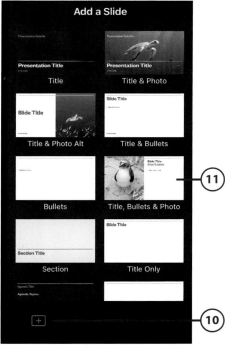

>>>*Go Further*

KEYNOTE LIVE

You can use the Keynote app as a remote control to present slideshows on other Apple devices, including Macs and iPhones. To use Keynote Live, open your presentation, tap More, and then tap Use Keynote Live. Follow the onscreen instructions to pair your iPad to another Apple device via Wi-Fi or Bluetooth.

Discovering More Productivity Apps

As good (and as free) as Apple's productivity apps are, they're not the only such apps available for your iPad. Google, Microsoft, and other software publishers offer similar productivity apps for iPad users, all available from Apple's App Store.

Google Docs, Sheets, and Slides

Taken together, Apple's Pages, Numbers, and Keynote apps comprise what some call an *office suite*—a collection of apps that enable you to perform most office productivity functions. Apple's productivity apps are kind of an ad hoc or unofficial suite, compared to the more formal app suites from other companies.

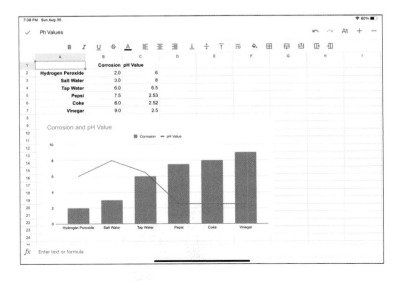

The Google Sheets spreadsheet app from the Google Docs suite

Case in point is Google's office suite, dubbed Google Docs. The apps in the Google Docs suite include Docs (word processing), Sheets (spreadsheet), and Slides (presentations). Unlike Apple's apps, Google's apps are cross-platform compatible; you can use the Google Docs apps on an iPad or iPhone, Android device, Windows computer, or Google Chromebook, if you have one.

All three Google Docs apps are free to download from the App Store and work great on the iPad.

Microsoft Excel, PowerPoint, and Word

Speaking of office suites, the big daddy of all office suites is Microsoft Office. Long available only on Windows PCs, Microsoft now has versions of its Office apps available for all platforms, including Apple's iPadOS. The apps include Word (word processing), Excel (spreadsheet), and PowerPoint (presentations). All of these apps are available for free download from the App Store.

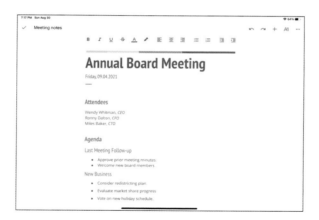

The Microsoft Word app from the Microsoft Office app

Microsoft's iPadOS versions of its Office apps aren't quite as robust as that of their PC counterparts; they don't offer all the functions or options available on the core Windows version. However, all document files are fully compatible between versions, so if you're used to using Office apps in your office, then using those same apps on your iPad makes a lot of sense.

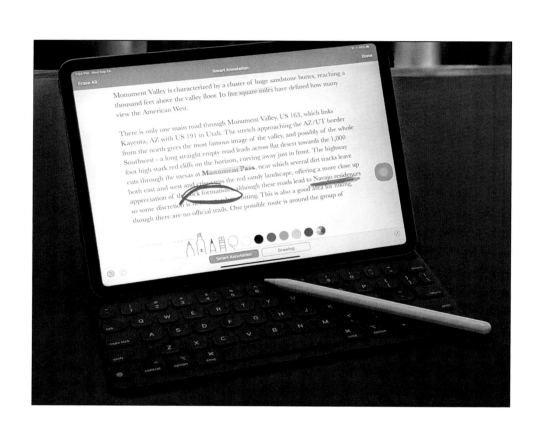

22

Using Pencils, Keyboards, and Trackpads

If you want to use your iPad for productivity, you may want to supplement the traditional touchscreen operation with alternative input devices. Apple and various third-party suppliers offer a variety of such devices—pencils and styli for drawing and writing, external keyboards for typing, and trackpads and mice for pointing and clicking.

Connecting Bluetooth Devices

All external pencils, styli, trackpads, and mice connect to your iPad via Bluetooth wireless technology. All non-Apple keyboards also connect via Bluetooth; Apple's Smart Keyboard and Magic Keyboard connect to iPad Pros via the built-in Smart Connector built in to those models.

Connect a New Bluetooth Device

All external Bluetooth devices connect to your iPad in the same fashion.

(1) Power on the external device and, if necessary, press the button for Bluetooth pairing.

(2) On your iPad, tap the Settings icon to open the Settings screen.

(3) Tap Bluetooth in the left column.

(4) Make sure Bluetooth is switched on.

(5) Your new device should appear in the Other Devices list in the right column. Tap to select it.

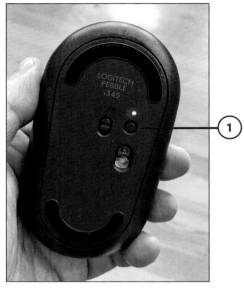

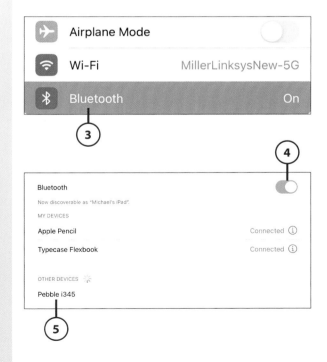

✈	Airplane Mode	⬤
🛜	Wi-Fi	MillerLinksysNew-5G
✳	Bluetooth	On

Bluetooth ⬤

Now discoverable as "Michael's iPad".

MY DEVICES

Apple Pencil Connected ⓘ

Typecase Flexbook Connected ⓘ

OTHER DEVICES

Pebble i345

6. When you see the Bluetooth Pairing Request, tap Pair.

7. The new device now appears in the My Devices list of paired devices.

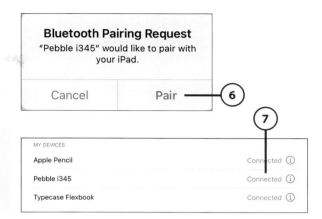

Bluetooth Pairing Request

"Pebble i345" would like to pair with your iPad.

Cancel   Pair — 6

MY DEVICES

Apple Pencil	Connected ⓘ
Pebble i345	Connected ⓘ
Typecase Flexbook	Connected ⓘ

Using Pencils and Styli

Pencils and other styli let you interact with the iPad screen without using your fingers and turn your iPad into a high-tech drawing tablet. They're popular with artists, illustrators, and engineers.

A pencil or stylus enables more precise onscreen drawing than just using your finger and often comes with enhanced functionality, such as highlighting or underlining onscreen items and navigating apps (by tapping with the stylus instead of your finger). You can also use a pencil to sign your name in online forms or write notes longhand rather than typing them.

There are many third-party pencils available for the iPad from companies such as Adonit, Logitech, and Wacom. Apple also offers its own pencil device, called the Apple Pencil. All of these devices connect to your iPad via Bluetooth. With the exception of the Apple Pencil, most styli are available for well under $100.

Work with the Apple Pencil

The Apple Pencil looks kind of like a fat white pencil, although it's really a stylus for touchscreen devices such as the iPad. Apple offers two variations of the Apple Pencil, the first-generation model ($99) and the second-generation model ($129). Each works similarly but with different iPad models, as detailed in Table 22.1.

Table 22.1 Apple Pencil Compatibility

Apple Pencil Model	Works With	Price
Apple Pencil (first generation)	iPad (6th, 7th, and 8th generations) iPad Air (3rd generation) iPad mini (5th generation) iPad Pro 9.7-inch iPad Pro 10.5-inch iPad Pro 12.9-inch (1st and 2nd generations)	$99
Apple Pencil (second generation)	iPad Air (4th generation) iPad Pro 11-inch (1st and 2nd generations) iPad Pro 12.9-inch (3rd and 4th generations)	$129

Note that older iPads, iPad Airs, and iPad minis do not work with either generation of Apple Pencil.

Using the Apple Pencil to write onscreen

Both the first- and second-generation Apple Pencils work in pretty much the same fashion. For those apps that support the Pencil, you can write, draw, and mark things up onscreen. The Apple Pencil is both pressure-sensitive and tilt-sensitive.

>>>Go Further

CHARGING THE APPLE PENCIL

All external devices have built-in batteries that periodically need to be charged. How you charge the Apple Pencil depends on the version you have.

To charge the first-generation Apple Pencil, remove the cap at the top end. This exposes a Lightning connector. Plug this connector into your iPad's Lightning port (on the bottom) and charging will commence.

The second-generation Apple Pencil magnetically attached to an iPad Pro

The second-generation Apple Pencil is even easier to charge thanks to the magnetic connector on the side of the iPad Pro. Just attach the Apple Pencil to the thin magnetic strip on the top (when held horizontally) or right side (held vertically) of the device, and it starts to charge.

Annotate a Document

A pencil or stylus is a great tool to use with a word processor or notes app, such as Apple Pages. You can use the tool to mark up and annotate your documents.

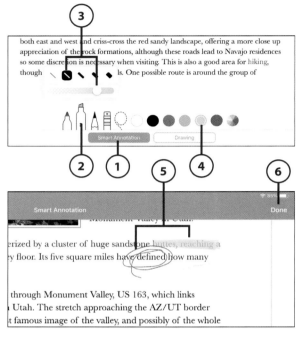

(1) Within a word processor or notes app, tap the screen with your pencil or stylus to activate your device. Select Smart Annotation.

(2) Tap to select a drawing tool.

(3) Tap and hold a tool to select size and opacity.

(4) Tap the desired color on the right.

(5) Start drawing or annotating.

(6) Tap Done at the top right to exit Smart Annotation mode.

Fill In Forms

The Apple Pencil can now be used to fill out text fields in forms, either in documents or web pages. Your hand-written text is automatically converted to typed text thanks to the new Scribble feature in iPadOS 14.

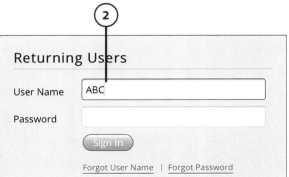

(1) Use the Apple Pencil to write within the text field.

(2) Your writing is automatically con-verted to typed text.

>>>Go Further

NAVIGATE iPAD SCREENS

You can also use a pencil or stylus to navigate your iPad. Instead of using your finger on the touchscreen, use the pencil or stylus to tap and press and swipe. It's easy and often more precise than tapping the screen with a fat finger (like mine!).

Draw Onscreen

Many illustration and presentation apps, such as Apple's Keynote or Adobe Sketch, support drawing with pencils and styli.

(1) Within the app, tap the screen with your pencil or stylus to enter Drawing mode. The palette appears at the bottom of the screen.

(2) Tap to select a tool.

(3) Tap the color icon to choose a color.

(4) Start drawing.

(5) Tap Done to exit Drawing mode.

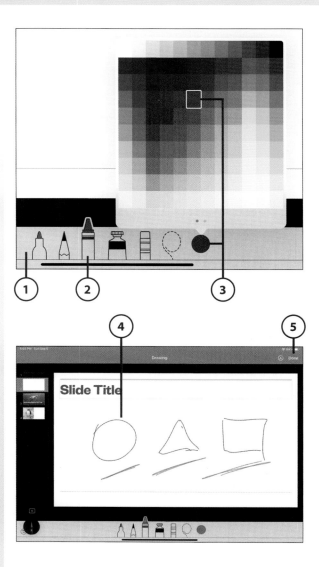

>>>*Go Further*

ADVANCED DRAWING PALETTES

Some drawing apps offer built-in palettes that you access with the Apple Pencil. The drawing tools in these programs are often more sophisticated than the Pencil's default tools.

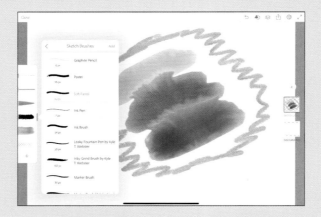

The drawing tool palette in Adobe Sketch

For example, Adobe Sketch places its palette on the left side of the screen and offers a selection of advanced drawing tools. Double-tap a tool to change its color; press a tool to select brush width and type.

Using External Keyboards

Apple and other companies make a plethora of external keyboards you can use with your iPad. Most of the keyboards connect to your iPad via Bluetooth and feature bigger keys with a more tactile feel than what you get with the iPad's onscreen keyboard.

Apple's two main keyboards are the Smart Keyboard and Magic Keyboard. Both let you position your iPad on an angle above the keyboard so that it all looks and works a little like a notebook computer. The Smart Keyboard is available in different sizes for different iPad models, including all the iPad Pros and the newest seventh-generation iPad and third-generation iPad Air. The Magic Keyboard is available for iPad Pro models only.

Connecting and Pairing the Smart Keyboard and Magic Keyboard

Both the Smart Keyboard and Magic Keyboard connect to your iPad magnetically. Compatible models have what Apple calls a Smart Connector on the back of the unit. When you attach the keyboard, it "snaps" on magnetically and automatically pairs with your iPad. When the Smart Keyboard or Magic Keyboard is attached, it starts working in place of the onscreen keyboard. Remove the keyboard, and your iPad reverts to the normal onscreen keyboard.

A Smart Keyboard connected to an iPad Pro

As far as charging Apple's keyboard, don't worry about it. Both the Smart Keyboard and Magic Keyboard charge automatically whenever they're attached to your iPad. It's all thanks to the Smart Connector, which transfers both data and power between your iPad and the keyboard. (Third-party keyboards are battery operated; you have to replace the batteries when necessary.)

Working with an External Keyboard

Working with any external keyboard is as simple as tapping the right keys. Most keyboards feature the same keys you find on a traditional Apple keyboard, including dedicated Control, Command (Cmd), and Option keys.

The keys on an Apple Keyboard

Using Trackpads and Mice

First introduced in iPadOS 13.6, iPads can now be controlled by trackpads and mice. When combined with an external keyboard, this lets you use your iPad just as you'd use a traditional notebook computer.

Logitech's Pebble Bluetooth mouse for the iPad

Some external keyboards, such as the Typecase Flexbook Touch shown here, incorporate touchpads as part of the keyboard, just as you'd find on a notebook computer. This type of multifunction accessory makes your iPad or iPad Pro virtually indistinguishable from a compact notebook computer.

An external keyboard with built-in trackpad

When you're using a trackpad or mouse, a round cursor appears onscreen. You move this cursor by moving the mouse or sliding your finger across the trackpad.

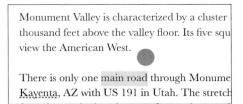

The onscreen cursor

Bluetooth Mice

If you're shopping for an external mouse for your iPad, make sure you get a Bluetooth mouse. Mice that have their own wireless transmitters and USB receivers, which are common in the world of personal computing, will not work with your iPad. The only way to connect a mouse to your iPad is via Bluetooth.

Configure Your Trackpad or Mouse

There are several options you can configure to personalize the way your trackpad or mouse works.

1. Tap the Settings icon to open the Settings app.
2. Tap General in the left column.
3. Tap Trackpad & Mouse.
4. Drag the Tracking Speed slider to adjust the speed of the onscreen cursor.
5. By default, your iPad uses natural scrolling, which mimics the movement of your fingers on the iPad screen. To disable this functionality, switch "off" the Natural Scrolling switch.
6. By default, the right button on a mouse is configured as a secondary click. If you're left-handed, you may want to reverse this; tap Secondary Click and then select Left, instead.

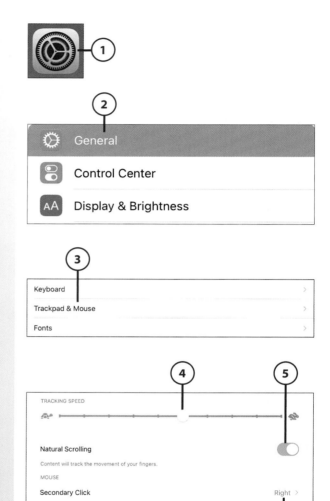

Use a Trackpad or Mouse

Operating your iPad with a trackpad or mouse is just like using one of those input devices on a personal computer. Tables 22.2 and 22.3 detail some of the more common actions you can employ.

Table 22.2   Common Trackpad Actions for the iPad

Action	Gesture
Click or select	Tap with one finger
Display App Switcher	Swipe down with three fingers
Display Control Center	Swipe one finger to the upper right until the cursor moves beyond the edge of the screen
Display Dock	Swipe one finger down until the cursor hits the bottom of the screen
Display Home screen	Swipe one finger down until the cursor goes beyond the bottom of the screen
Display Notification Center	Swipe one finger to the top or upper left until the cursor moves beyond the edge of the screen
Scroll	Use two fingers on the trackpad
Switch between apps	Swipe up with three fingers
Zoom in	Expand fingers out
Zoom out	Pinch fingers in

Table 22.3   Common Mouse Actions for the iPad

Action	Operation
Click or select	Press left mouse button
Click and hold	Press and hold left mouse button
Display App Switcher	From any Home screen, drag cursor beyond the bottom of the screen
Display Control Center	Click the status icons at the top-right corner of the screen
Display Dock	Drag cursor to the bottom of the screen
Display Notification Center	Click the status icons at the top-left corner of the screen
Display Search panel	From any Home screen, rotate the mouse's scroll button downward
Display Slide Over app	Drag cursor beyond the right edge of the screen
Drag	Click and hold item with left button; then move the mouse

In this chapter, you learn how to find and play games on your iPad.

→ Finding Games to Play
→ Discovering Popular Games
→ Playing Games with Apple Arcade

Playing Games

Yes, you use your iPad to send and receive emails and text messages, listen to music, watch videos, and read and post to Facebook and other social networks. You may even take a few photos or videos or do some real productive work.

But let's be honest. You also may spend a fair amount of time on your iPad playing games. And if you don't, your kids or grandkids do. No need to hide it; the iPad is a great game-playing device, and there are lots of fun games you can play on it.

Finding Games to Play

A game is just a certain type of app. Just as you find regular apps for your iPad in Apple's App Store, that's also where you find the games you want to play on your iPad.

Download Games from the App Store

The App Store has a section just for games. Some of these games cost money, but many are free. It's easy to browse through the game categories or search for specific games to download.

(1) From your iPad's Home screen, tap the App Store icon to open the App Store.

(2) Tap the Search icon to search for a specific game. *Or...*

(3) Tap the Games icon to view all games.

(4) Scroll down the page to see recommended games by type.

(5) To view games by category, scroll to the Top Categories section and tap See All.

(6) Tap the type of game you want to browse: Action, Board, Casual, and so forth.

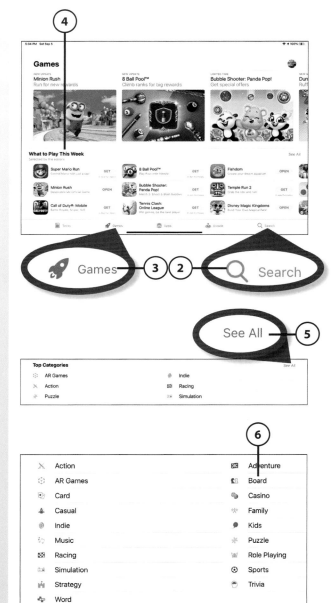

(7) Tap the game you want to play. You see a detailed page for that game, which includes user reviews.

(8) Some games are free. To download a free game, tap Get.

(9) Many games cost money to play. To purchase a game, tap the price button.

(10) Follow the onscreen instructions to complete the purchase on your device.

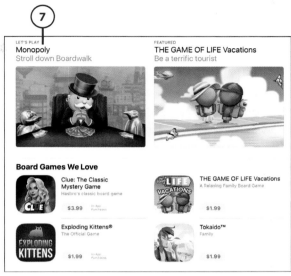

Discovering Popular Games

What types of games can you find in the App Store? Apple organizes games by category, as you'll soon discover.

Action Games

Playing the Super Mario Run action game

Action games are those that emphasize physical challenges and require good hand-eye coordination. Popular action games for the iPad include Angry Birds 2 and Angry Birds Friends, Call of Duty: Mobile, Cube Surfer, Fruit Ninja, Sonic Dash, Super Mario Run, and Subway Surfers.

Adventure Games

Playing the Minecraft adventure game

When you play an adventure game, you assume a starring role in some sort of episodic story. Popular adventure games include Grand Theft Auto: San Andreas, LEGO Jurassic World, Terraria, Township, and Zooba.

All that said, the most popular adventure games for the iPad or any device are Minecraft and Roblox. These are world-building games, where you use your block-like characters to explore new environments, find resources, and build complex structures. Although these games might not be your thing, your grade school–aged children and grandchildren are probably already hooked.

AR Games

Playing the Pokémon GO AR game

AR stands for *augmented reality*. AR games integrate computer-generated characters with your real-world environment on your iPad screen in real time.

Popular AR games on the iPad include AR Dragon, Avo!, CSR Racing 2, Jurassic World Alive, Harry Potter: Wizards Unite, and Lightstream Racer. Arguably the most popular AR game is Pokémon GO, which lets you collect Pokémon as you walk around your neighborhood or town.

Board Games

Playing The Game of Life board game

Board games for the iPad are just like traditional board games, except you play them on your iPad screen, either by yourself (against the computer) or with others (over the Internet). Popular board games include Battleship, Clue, The Game of Life, Monopoly, Risk, and Roll for the Galaxy.

Card Games

Playing the UNO card game

There are many games for the iPad that mimic traditional physical card games. These include canasta, gin rummy, Skip-Bo, solitaire, and UNO.

Casino Games

Playing Cash Casino Slots casino game

Just about any game you can play in a casino is available in an iPad-friendly version. Popular casino games include variations of bingo, blackjack, craps, poker, and slots—as well as collections of multiple games, just like a real casino.

Casual Games

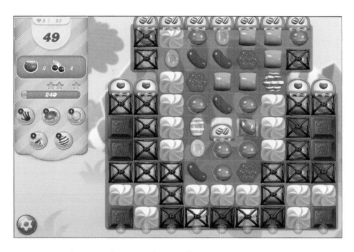

Playing the Candy Crush Saga casual game

Casual games are fun and easy to play. When you just want to waste a few minutes between things or unwind a little, a quick casual game is the way to go.

The most popular casual games in the App Store include the various Candy Crush Saga games, Fruit Ninja Classic, Geometry Dash, Sky Roller, and Subway Surfers.

Family Games

Playing the Disney Magic Kingdoms family game

Playing games with the whole family is a fun thing to do. In addition to board and card games, some of the most popular family games in the App Store include Animal Crossing: Pocket Camp, Disney Magic Kingdoms, Minion Rush, PJ Masks: Racing Heroes, SpongeBob: Sponge on the Run, as well as arcade classics such as PAC-MAN and Ms. PAC-MAN.

Indie Games

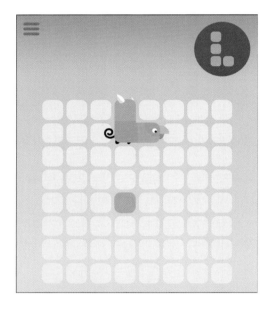

Playing the Blocki! indie game

These are games from independent game developers (instead of the big game companies). Check here for some of the newer and more innovative games of all types, including Blocki!, Dissembler, High Rise, Lost Spaceships, and Shadow Frog.

Kids Games

Playing the Crayola Create and Play kids game

When you want to occupy the youngest members of your family, check out these fun kid-friendly games. The most popular include Barbie Dreamhouse Adventures, Crayola Create and Play, Disney Coloring World, Miga Town, Toca Life, and World of Peppa Pig.

Also good are educational apps such as ABCmouse.com, Epic—Kids' Books & Reading, GoNoodle Kids Videos, Kiddopia—ABC Toddler Games, and PBS KIDS Games.

Music Games

Playing SongPop2—Guess the Song

Many games revolve around music creation and quizzes. In particular, check out Game of Songs—Music Gamehub, Magic Tiles 3: Piano Game, Marshmello Music Dance, SongPop 2—Guess the Song, and Tiles Hop—EDM Rush.

Puzzle Games

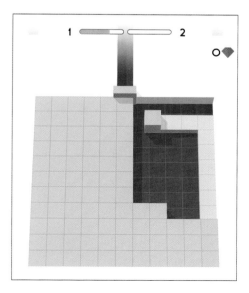

Solving the Color Fill 3D puzzle game

These games require you to move tiles or other objects around, connect dots together, or otherwise solve various sorts of puzzles. Popular puzzle games for the iPad include Brain Wash—Puzzle Mind Game, Color Fill 3D, Emoji Puzzle!, Gardenscapes, and Match 3D.

Racing Games

Racing an exotic car in the Asphalt 9: Legends racing game

Strap yourself in to race all sorts of virtual vehicles in Asphalt 9: Legends, Fast Driver 3D, Forza Street: Tap to Race, Mario Kart Tour, Speed Moto Dash: Real Simulator, Spiral Roll, and Wheel Race.

Role-Playing Games

Playing the Injustice 2 role-playing game

Role-playing games are even more involved adventure games in which you take on the role of a specific player. Popular games of this type include Drive Thru 3D, Gacha Club, Injustice 2, MARVEL Future Fight, Rise of Kingdoms, Star Wars: Galaxy of Heroes, and Surgeon Simulator.

Simulation Games

Playing the Township simulator

When you want to simulate some fun real-world activity or persona, check out games like Fallout Shelter, Plague Inc., Pocket Build, SimCity Buildit, Stardew Valley, Township, and Ultimate Jungle Simulator.

Sports Games

Playing the Madden NFL 21 Mobile Football game

Just about every popular sport is represented in the App Store, including Bouncy Hoops, Golf Battle, Madden NFL 21 Mobile Football, NBA 2K Mobile Basketball, and R.B.I. Baseball 20.

Strategy Games

Playing the Draw Defence strategy game

Strategy games require more thinking than action. The most popular include Clash of Clans, Draw Defence, Five Nights at Freddy's, Magic: ManaStrike, State of Survival: Zombie War, and West Game.

Trivia Games

Playing the Trivia Crack trivia game

Who was the first actor to play James Bond? What is the longest river on Earth? What was the top boy band in the 1990s? Answer these and other questions in games such as 100 PICS Quiz—Picture Trivia, Psych! Outwit Your Friends, QuizUp, SongPop 2—Guess the Song, Trivia Crack, and Trivia Star: Trivia Games Quiz.

Word Games

Playing the Word Cookies word game

For many people, I've saved the best category for last. Word games let you put your word-building skills to the test, playing either solo or against other players. The most popular word games for the iPad include Geekwords Daily Crosswords, Scrabble GO—New Word Game, Toliti, Typoman Mobile, Word Collect: Word Games, Word Cookies!, Word Domination, Words with Friends—Word Game, and Wordscapes.

It's Not All Good

In-Game Purchases

Many games often ask—again and again—that you pay money to get additional levels or resources. These in-game purchases can quickly add up, so think twice before tapping the "buy now" button in any given game.

Playing Games with Apple Arcade

In addition to offering individual games in the App Store, Apple offers a sub-scription games service called Apple Arcade. When you subscribe to Apple Arcade, you get access to more than 100 games every month. You can play any of the games in Apple Arcade as much as you want, wherever you are. After you download a game, you don't need an Internet connection to play.

Your Apple Arcade subscription works on any Apple device—iPad, iPhone, Mac, or Apple TV. Your subscription is good for up to six family members, too.

Apple Arcade costs $4.99 per month, and you sign up from the App Store app. Just select the Arcade tab and follow the onscreen instructions to purchase your subscription.

Finding Apple Arcade Games

Browsing and searching for games in Apple Arcade is just like finding games in the App Store.

(1) From within the App Store app, tap to select the Arcade tab.

(2) Scroll down to view recommended games—Popular Arcade Games, Arcade Games for You, New Arcade Games, and the like.

(3) Tap See All to view all games in that category.

Downloading Apple Arcade Games

Once you've paid for the subscription, downloading games from Apple Arcade is always free.

(1) Tap Get to download a game. *Or…*

(2) Tap a game to view more details.

(3) Read more about the game and then tap Get to download it.

(4) The game is installed on your iPad. Tap the thumbnail to launch the game.

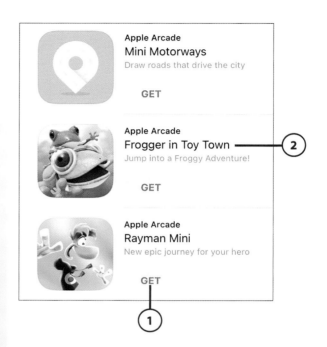

 Start playing!

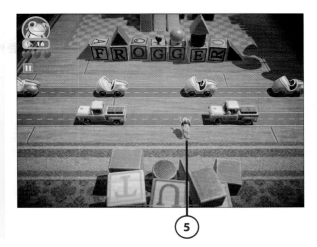

5

AARP Games

The AARP website is another good source for online games you can play on your iPad via Safari or another web browser. AARP Games includes dozens of fun games, including Backgammon, Chess, Classic Solitaire, Daily Crossword, Jigsaw (puzzles), Mahjongg Remix, Scramble Words, and Sudoku.

Many of these games are free for all to play. Others are reserved for AARP members. (AARP membership is just $16 per year, and well worth it for these and other benefits.) See what's available at games.aarp.org.

In this chapter, you learn how to manage the files stored on your iPad.

24

Managing Files on Your iPad and in the Cloud

How do you get your photos and other files from your iPad to your computer—or vice versa? How do you manage those files stored on your iPad? How can you share your files with others? This chapter shows you how to do all these things.

Managing Your iPad's Storage

First off, I want to show you how to find out how much storage you're using on your iPad. You have limited storage, after all; if you fill it, you won't have enough space to install new apps or shoot new pictures or videos.

Examine Available Storage

You can see how much storage space you've used—and still have available—from your iPad's Settings screen.

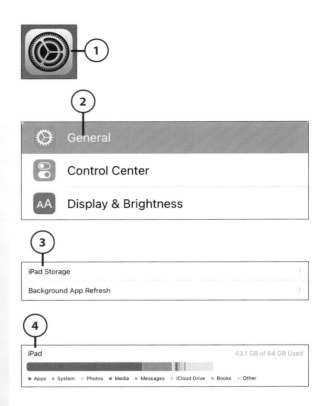

1. Tap the Settings icon to open the Settings screen.

2. Tap General in the left column.

3. Tap iPad Storage in the right column.

4. The iPad bar graph shows the amount of storage space used by type of file (Media, Apps, Photos, and so on). The lightest gray area shows the amount of remaining storage space available.

Manage Available Storage Space

Which apps are using the most storage space? If you're running short on space, which apps are the best to delete?

Fortunately, Apple makes it easy to look at storage on an app-by-app basis—and, just as easily, delete those apps and files that you're not using.

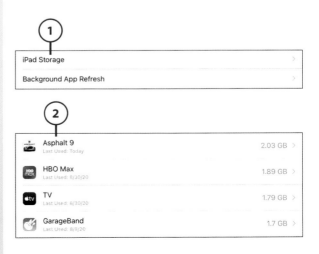

1. From the Settings screen, tap General and then tap iPad Storage.

2. At the bottom of the iPad Storage screen, you see a list of all the apps installed on your iPad, with the largest apps (those using the most storage space) first. Tap an app to view more details.

3 To delete the app but keep its data and files, tap Offload App.

4 To completely delete the app and all its data and files, tap Delete App.

It's Not All Good

You Can't Delete Everything

You can delete all third-party apps that you install, but you can't delete some of Apple's built-in apps—even if you never use them.

>>>Go Further

WHAT TO DELETE

When you're looking to free up storage space on your iPad, it makes sense to look at the largest apps first—which is why Apple sorts them that way. You get the most bang for your buck by deleting a single large app than by deleting several smaller ones. So look at the apps in order of largest to smallest and find those you no longer use or want. The first ones you come across are the ones you should delete.

Working with the Files App

The Files app lets you manage files and documents stored on your iPad, as well as those stored in the iCloud cloud storage system.

View and Open Files

You use the Files app to view and manage files you've downloaded or copied to your iPad.

1. From your iPad's Home screen, tap the Files icon to open the Files app.

2. You can view files stored on your iPad, on your iCloud Drive, or on any external storage device connected to your iPad. To view your iPad's files, tap On My iPad in the Locations section on the left.

3. You can view files in three different views—Icons, List, or Columns. Tap the Views icon to make a selection. (The Views icon changes depending on the view currently selected.)

4. Tap Icons to view files as icons or thumbnails (for images).

5. Tap List to view a list of files.

6. Tap Columns to view files and folders in a column format.

7. Sort folders by name, date, size, kind, or tags by tapping to select how you want your files and folders sorted.

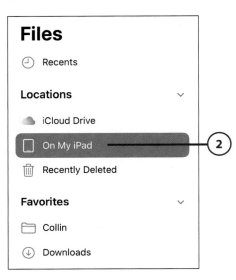

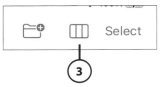

8 Files are stored in folders. In the column layout, tap a folder to view its files in the next column.

9 Tap a file to view information about that file in a further column.

10 Tap Open to open the file.

Work with Files

The Files app lets you share, duplicate, move, delete, or copy files.

1 From within the Files app, tap Select.

2 Tap to select one or more items. A white check mark in a blue circle indicates selected items.

3 Tap Share to share the selected files with another user; then select how and with whom you want to share.

4 Tap Duplicate to create duplicates of the selected files (in the same folder).

5 Move the selected files to another location by tapping Move; then select another folder.

6 Delete the selected files by tapping Delete.

7 Tap More and then tap Copy to create copies of the selected files in another location. Then navigate to another folder, press and hold within that folder, and tap Paste.

8 Tap Done when you're done working with these files.

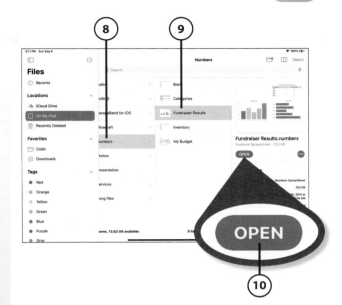

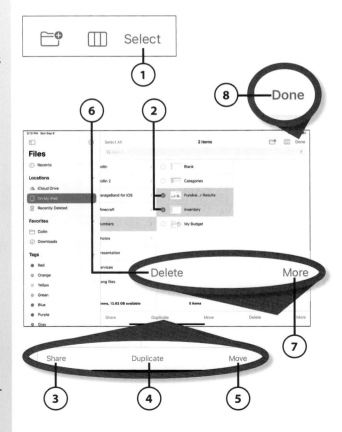

Create a New Folder

You don't have to settle for the default folders created by your iPad apps. You can create your own custom folders and move or copy appropriate files there.

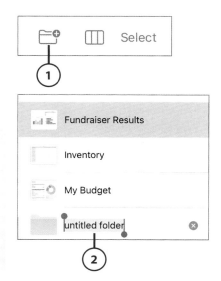

(1) Navigate to a specific location and then tap the New Folder icon.

(2) Type a name for this folder.

Using External Storage

Another new feature of iPadOS 13 is the ability to use external storage devices with your iPad. This might be a USB flash drive or an external hard drive. All you have to do is connect the external drive to your iPad via the iPad's Lightning or USB-C connector.

Make the Connection

The most difficult part of using an external storage device is connecting it to your iPad. Most storage devices connect via the traditional USB 1.0 or 2.0 Type-A connector. Your iPad doesn't have one of these. You need to use either a USB Type-A to Lightning cable or connector or a USB Type-A to Type-C cable or connector—the specific type depends on the type of connector on the bottom of your iPad.

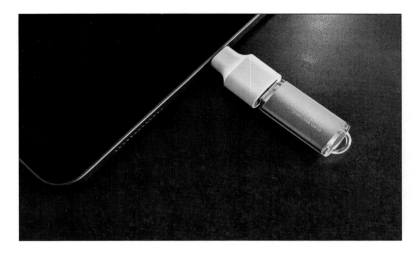

A USB flash drive connected to a USB Type-A to Type-C connector, connected to an iPad Pro

From there, it's a simple matter of connecting your external storage device to the cable or connector.

Work with Files

Once the external device is connected to your iPad, it's recognized and displayed in the Location section of the Files app.

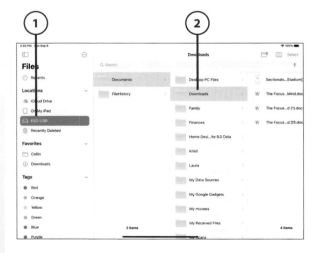

(1) From within the Files app, tap the name of your external device in the location section.

(2) Tap to navigate to the folder and file you want.

3 Tap a file to view its details.

4 Tap Open to open the file. *Or…*

5 Tap Select to select one or more files.

6 Tap each file you want to select.

7 Tap to Share, Duplicate, Move, Delete, or Copy the selected file(s) either to another folder on your external drive, to your iPad, or to your iCloud storage (discussed next).

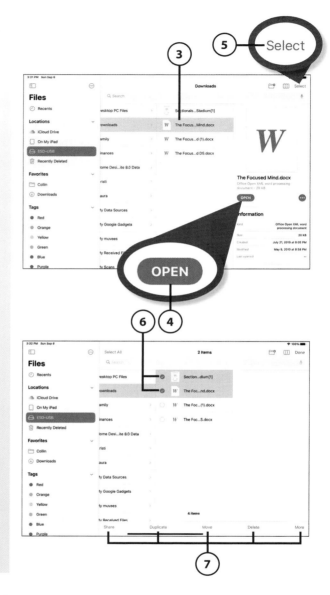

Working with iCloud

Apple's iCloud is a cloud-based storage service, which means it stores your files on Apple's servers, which you access over the Internet. All files are synced wirelessly when you're connected to a Wi-Fi network, so you don't have to physically connect your iPad to your computer. By default, all files you create with the Pages, Numbers, and Keynote apps are stored in your iCloud storage.

>>>*Go Further*

UNDERSTANDING iCLOUD AND CLOUD STORAGE

When it comes to storing data, there are two types of storage: local and cloud. Local storage means that your data is stored locally, on your device (in your case, your iPad). Cloud storage stores your data on computer servers in the "cloud"—that is, on the Internet. The advantage of local storage is that you always have your data at hand, even if you don't have an Internet connection. The advantage of cloud storage is that you can access your data from any device connected to the Internet from wherever you happen to be. In addition, should something happen to your iPad, you can still access the data. All you need is an Internet connection and the username and password for your cloud storage account.

Apple's cloud storage service is called iCloud, and it's integrated into iPadOS and your iPad. This is a good thing because you have a lot more storage space available in the cloud than you do on your iPad. You also can access the photos and videos you shoot with your iPad from any other device—your iPhone, for example, or your desktop or laptop computer.

One nice thing about how iCloud works is that once you enable it, you really don't have to do much else. Any new photo or file you create on your iPad is automatically uploaded to iCloud; you don't have to manually transfer any files. Your data also syncs between multiple devices; if you have a shared calendar, when you add an event on your iPad's Calendar app, that new event appears on all other devices linked to your iCloud account.

Access isn't limited to Apple devices. You can log on to iCloud.com from any computer or mobile device by using any web browser. It's all connected.

You were prompted to create an iCloud account when you first powered up your new iPad. If you didn't create an account then, you can create a new account at some later time by going to iCloud.com and selecting the Create Apple ID option.

You get 5GB of free storage with your iCloud account. That's a lot, and it might be all you need. If it's not enough storage space, however, you can upgrade your account to get more storage. In the United States, you pay $0.99 per month for 50GB of storage (that's 10 times the default storage space), $2.99 per month for 200GB, or $9.99 per month for a whopping 2TB. (That's two terabytes, or 2,000 gigabytes!)

Configure iCloud on Your iPad

You probably set up iCloud when you first powered up your iPad. You can change any iCloud settings at any time from your iPad's Settings screen.

(1) From your iPad's Settings screen, tap your account name or picture in the left column.

(2) In the right column, tap iCloud.

(3) The Storage section tells you how much iCloud storage space you have free.

(4) Tap Manage Storage to manage your storage or buy more storage space.

(5) iCloud storage is enabled by default. To turn it off for all apps, scroll down and tap "off" the iCloud Drive switch.

(6) To enable or disable iPad storage for any specific app, tap "on" or "off" the switch next to any app.

(7) Tap Photos to manage photo sharing via iCloud.

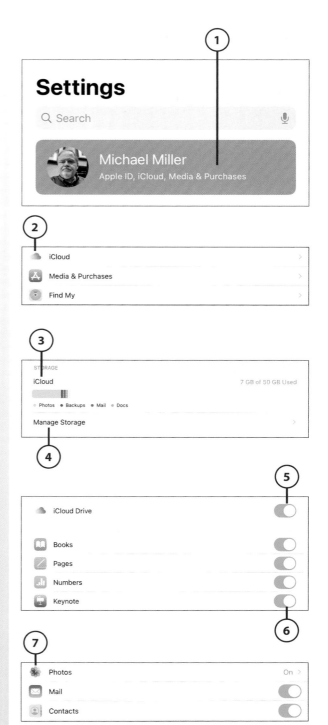

8. Tap "on" the iCloud Photos switch to automatically upload and store your photo library in iCloud. (This is enabled by default; tap it "off" to disable this feature.)

9. Tap Optimize iPad Storage to store all your photos online and automatically delete them from your iPad. *Or…*

10. Tap Download and Keep Originals to keep your original photos on your iPad (and copies in iCloud).

11. Tap "on" the Shared Albums switch to create photo albums to share with other users.

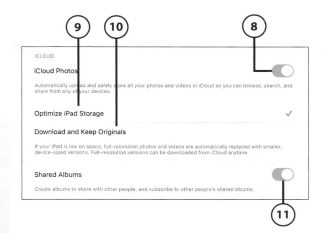

Back Up Your iPad to iCloud

The data that you store on your iPad—pictures, videos, music, and more—is valuable to you. You don't want to lose these items if you happen to lose or damage your iPad or if your iPad quits working.

This is why you want to back up all your data—and settings and apps—in case something bad happens. When you have a backup of your important stuff, you can then restore those items to your iPad when it's found or fixed.

You back up your iPad files online to your iCloud account. It's easy to do.

When your iPad is connected to the Internet via Wi-Fi, backups happen automatically in the background. All you have to do is configure your iPad for automatic iCloud backup.

1. From your iPad's Settings screen, tap your account name or picture in the left column and iCloud in the right column.

2. Scroll down the list of apps and tap iCloud Backup.

3. Tap "on" the iCloud Backup switch. (This is probably enabled by default.)

4. If you want to back up your data now, tap Back Up Now.

>>>Go Further
ACCESS iCLOUD FROM YOUR COMPUTER

You can view all items stored in your iCloud account from any connected device, including your desktop or laptop computer. Just open your web browser, go to iCloud.com, and then enter your Apple ID and password.

Once you've signed in, you see icons for all the iCloud-related services—Mail, Contacts, Calendar, Photos, and such. If you want to view your uploaded photos, click Photos. If you want to view other uploaded files, click iCloud Drive. From there, you can navigate through the folders to view individual files.

Using AirDrop and AirPlay to Share with Other Devices

One nice thing about committing to the Apple ecosystem (that is, using multiple Apple devices—iPads, iPhones, Mac computers, and the Apple TV set-top device) is that it's easy to share data and files between each device. This is particularly easy using Apple's AirDrop and AirPlay feature. AirDrop lets you easily sync files between devices, whereas AirPlay enables wireless streaming of music and videos to other devices.

Share Files with AirDrop

Use AirDrop to share photos, videos, locations, websites, and more with other compatible devices. AirDrop uses both Wi-Fi and Bluetooth to transfer data; Bluetooth is used to locate a nearby Apple device, and then the devices create a unique Wi-Fi connection between each other.

To use AirDrop, you must have your iPad's Wi-Fi and Bluetooth turned on. All transfers are encrypted for security. You also need to turn on AirDrop. To do this, open the Settings app, select General, and then tap AirDrop. You can choose to share with Everyone or with Contacts Only. You can also turn off AirDrop receiving if and when you'd rather not receive files from others.

Here's how to share a file with AirDrop.

1. Within an iPad app, tap the Share icon. (In some apps, you may have to tap More and then tap Share.)

2. Tap the AirDrop icon.

3. Nearby devices are displayed. Tap the name of a device to share this file with that user.

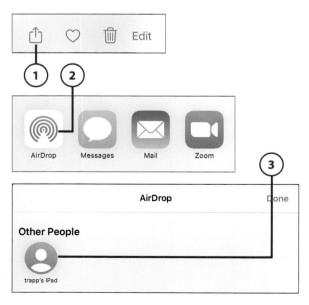

4. The other user receives an AirDrop sharing request. They should tap Accept to receive the file.

Stream Media with AirPlay

Use AirPlay to stream music and videos from your iPad to any compatible device in real time. You can use AirPlay to stream music from your iPad to an Apple TV box, AirPort Express device, AirPlay-enabled wireless speakers, or other AirPlay-enabled devices (such as some audio/video receivers). You can also use AirPlay to stream photos and videos to an Apple TV box.

To stream media via AirPlay, both your iPad and the target device need to be connected to the same Wi-Fi network. You may also need to enable AirPlay on the other device.

1. With music or a video playing, swipe down diagonally from the top-right corner of the screen to open the Control Center, and then long press the Now Playing control.

2. From the enlarged Now Playing control, tap the AirPlay icon.

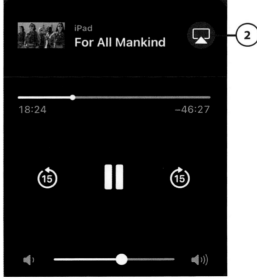

3 AirPlay looks for and displays nearby devices. Tap the device to which you want to stream. The current media is now streamed to the selected device.

4 Tap iPad to return playback to your iPad.

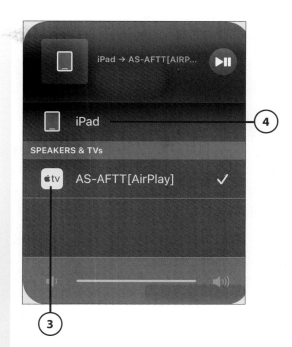

Screen Mirroring

Your iPad also offers Screen Mirroring, which lets you display your entire iPad screen (apps and all) on an Apple TV set-top device. Open the Control Center, tap the Screen Mirroring tile, and then select your Apple TV device from the list. (If a passcode appears on your TV screen, enter that code into the appropriate field on your iPad screen.) You now see your iPad screen on your TV display.

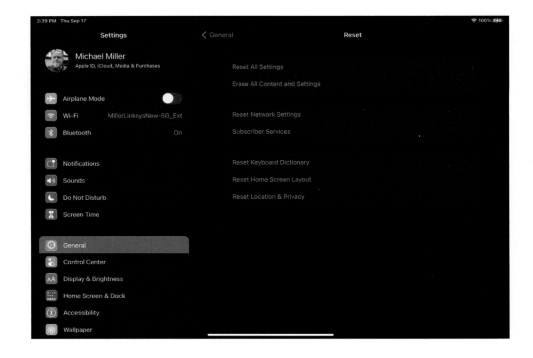

In this chapter, you learn how to troubleshoot and fix problems you may encounter with your iPad.

→ Troubleshooting iPad Problems
→ Updating and Resetting Your iPad

Fixing Common Problems

Something wrong with your iPad? Maybe it's just running slowly or having trouble connecting to the Internet. Have no fear—most iPad problems are easy to fix, if you know what to do.

That's where this chapter comes in. Read on to learn how to troubleshoot and fix the most common problems you might encounter with your iPad.

Troubleshooting iPad Problems

Your iPad is a technological marvel. The amount of computing power packed into such a small device is nothing short of amazing. When I was growing up in the 1960s, before the dawn of personal computers, video games, and even handheld calculators, we couldn't have conceived of such a device or the things it could do.

That said, with all technology comes some degree of complexity—and with complexity sometimes comes problems. Have you ever picked up your iPad and not been able to turn it on? Or has the screen frozen on you? Or has the darned thing dropped an Internet connection? Then you know what I mean. The iPad is great, but it ain't always perfect.

So when your iPad is less than perfect, work through the troubleshooting tips in this section. Chances are you'll have it up and running again in no time.

Your iPad Is Frozen

Okay. You're using your iPad, tapping along, when all of a sudden nothing you tap seems to work. You can do all the tapping in the world, but your screen is completely frozen. What do you do?

Well, the one thing you don't want to do is panic. That's because a frozen iPad, as frustrating as it may seem, typically isn't a major problem. There are a few simple things to try that will likely get your iPad working again.

1. Try restarting your iPad by pressing and holding the On/Off button for a few seconds. (On an iPad Pro, press and hold both the On/Off and either Volume button.) When the slider appears, slide it to the right to power off. Then restart the iPad by pressing the On/Off button again until the Apple logo appears.

2. If your iPad won't restart, then you need to do what Apple calls a *force restart*. Don't worry; this won't delete any of your content. Press and hold the On/Off and Home buttons at the same time, for at least 10 seconds, until you see the Apple logo appear onscreen. When you see the logo, your device has restarted and should be unfrozen.

3. If you have a model without a Home button, like an iPad Pro, then the force restart process is slightly different. It's a three-step process: First, press and release the Volume Up button. Second, press and release the Volume Down button. Third, press and *hold* the On/Off button until you see the Apple logo appear onscreen. That should do it.

Your iPad Won't Turn On

What do you do if you can't even turn on your iPad? This situation may be caused by a few different problems.

1. Your iPad may essentially be frozen in sleep mode. Try resetting the device by pressing and holding the On/Off and Home buttons at the same time for at least 10 seconds. (On an iPad Pro, press and release the Volume Up button, press and release the Volume Down button, and then press and hold the On/Off button.) If you see the Apple logo onscreen, your iPad is good to go.

2. It's also possible that your iPad is simply out of juice. Plug it in and let it charge for 10 minutes or so, and then try starting it up again. (If you see the charging screen with a low battery graphic a few minutes after you plug it in, this was definitely your problem.)

3. If you don't see the charging screen when you plug in your iPad, check the connecting cable, power adapter, and the connections between the cable, connector, and wall outlet. Make sure everything is firmly connected, and that the power outlet actually has power. (That is, turn on the wall switch!)

It's Not All Good

Freezing During Startup

If your iPad starts up but then freezes during the process (the Apple logo appears but doesn't go away), you can try connecting your device to your computer, launching iTunes, and then doing a force restart. When you get to the Apple logo screen, keep pressing the On/Off and Home buttons until you see the Recovery Mode screen. From here, choose the Update option, and iTunes will try to reinstall iPadOS without deleting any of the data stored on your iPad.

Your iPad Won't Turn Off

If you think it's a pain when your iPad won't turn on, imagine what it's like when your device won't turn *off*. If this happens to you, try the following:

1. See if a given app is causing the problem. Swipe up and then to the right from the bottom of the screen or press the Home button twice to display

the App Switcher and then close each app individually (by pressing and dragging the app thumbnail off the top of the screen). When all the apps are closed, try turning off your iPad.

2. Do a force restart by holding the Top and Home buttons simultaneously for at least 10 seconds. (Or, on an iPad Pro, press and release the Volume Up button, press and release the Volume Down button, and then press and hold the On/Off button.)

3. If your iPad still won't shut off, just keep it running until the battery runs out; then recharge it.

An Individual App Freezes or Doesn't Work Right

If an individual app freezes on you, you have several options to try.

1. Swipe up and to the right from the bottom of the screen or press the Home button twice to display the App Switcher. Drag the unresponsive app up and off the screen. You can then relaunch the app and see if it works properly.

2. If the previous step didn't solve your problem, you might have to restart your iPad to close the frozen or misbehaving app. Press and hold the Top and Home buttons for about 10 seconds until the Apple logo appears onscreen. (On an iPad Pro, press and release the Volume Up button, press and release the Volume Down button, and then press and hold the On/Off button.)

3. If the app is still giving you trouble, uninstall and then reinstall the app. From the Home screen, press and hold the icon for the app until it starts to jiggle, and then tap the X to delete the app. You can press the Home button (or tap Done on the screen) to exit delete mode and then go to the App Store and reinstall a new version of the app.

Your iPad Runs Slowly

When your iPad seems slower than normal, it's probably because there are one or more apps running that are slowing things down. Here are a few things to try:

1. Swipe up and to the right from the bottom of the screen or press the Home button twice to display the App Switcher and then close any app that you're

not currently using. (If you have more than a half-dozen apps open, that's probably too many.)

2. Do a force restart by holding down the Top and Home buttons simultaneously for at least 10 seconds. (Or, on an iPad Pro, press and release the Volume Up button, press and release the Volume Down button, and then press and hold the On/Off button.)

Force Restart

As you might have gathered, doing a force restart is an amazing cure-all. You'd be surprised all the problems that this solves!

Your iPad's Wi-Fi Connection Doesn't Work

You're sitting in your living room or in a local coffeeshop, doing your daily Internet surfing, when all of a sudden you can't access any websites or send email. Something's happened to your Wi-Fi connection—but what?

1. Turn your Wi-Fi off and back on by opening the Control Center and tapping "off" the Wi-Fi control. Wait a few seconds, and then tap the control back on.

2. Make sure you're connected to the right network. Open the Settings page, select Wi-Fi, and then look at the list of available Wi-Fi networks. If the wrong network is selected, tap the right one to connect.

3. If that doesn't work, try restarting your iPad. Hold down the On/Off button (or, on an iPad Pro, hold down the On/Off button and either Volume button) and then slide the slider to the right to power off. You can restart your iPad normally and see if the Wi-Fi is now working.

4. You can also reset your iPad's network settings. From the Settings screen, select General, tap Reset, and tap Reset Network Settings. This also restarts your iPad; you'll have to go in and choose the correct Wi-Fi network again.

Your iPad Charges Slowly or Not at All

If you find your iPad taking longer than usual to charge, not charging at all, or simply not holding a charge as long as it used to, there are several things to check.

1. Make sure the connecting cable is firmly plugged into your iPad and to the power adapter. For that matter, make sure the power adapter is firmly plugged into a wall outlet—and that the wall outlet is turned on.

2. Try changing charging cables. Cables do go bad, and when they do, charging times suffer. (Or they just quit working altogether—which means you can't charge at all!)

3. If you have a spare power adapter around, try using that one instead of your old one. These gizmos go bad, too.

4. If you've been trying to charge your iPad by connecting to your computer (via USB), plug directly into the wall outlet (via the power adapter) instead. Charging via computer is much, much slower than charging via wall outlet. In fact, some computers simply don't send enough juice to their USB ports to charge your iPad at all!

It's Not All Good

Old Batteries

If your iPad isn't holding a charge as long as it used to, it could be that your device's battery is getting old. iPad batteries become less efficient over time; if you've had your iPad for a year or more, it simply won't hold as much of a charge as it did when new. The only solution to this is to install a new battery, which you can't do yourself. You'll have to go to an Apple Store or other service center for this.

Your iPad Doesn't Rotate

One of the things I like about the iPad is how the screen rotates when you turn the device one direction or another. What do you do if your iPad doesn't automatically rotate?

1. First, know that not all iPad apps rotate. (This is more of a problem with older apps than newer ones, although it's also a "feature" of some games.) If you're using an app and turn your iPad and nothing happens, switch to another app or back to the Home screen to see if it rotates. If so, then the problem is a non-rotating app.

2. It's possible that you accidentally turned on something called *rotation lock*. With rotation lock enabled, your iPad screen simply won't rotate no matter what you do. To switch this off, swipe diagonally down from the top corner of the screen to open the Control Center. If the rotation lock is enabled, the Lock Rotation icon is red against a white button background. Tap this button to turn it off. (You need to close the Control Center for the auto-rotation to start working again.)

Your iPad Keeps Asking for Your iCloud Password

Here's another thing that's happened both to me and to my wife (on different devices). I start using my iPad, and sooner or later (mostly sooner) it nags me with a prompt to sign in to my iCloud account. Even if I'm already signed in to my iCloud account, it just keeps nagging me.

Is there any way to stop this constant nagging? Here's how.

1. It may sound obvious, but try signing in to your iCloud account. Sometimes doing what your device asks for is the proper approach.

2. If the iPad keeps prompting you to sign in to your iCloud account, something's stuck in the authorization process that makes your iPad think it needs the password when it really doesn't. Try restarting your iPad and see if this makes the annoying prompts go away.

3. If this doesn't do it, try signing out of iCloud and then signing back in to the service. You do this from the Settings screen; tap your ID name or picture, then tap Sign Out. (You're prompted to enter your iCloud password to sign out.) After you sign out, return to the same Settings screen and sign back in again.

>>>Go Further
FORGOTTEN PASSWORD

What do you do if you simply can't remember your iCloud password? You need to reset your account with a new (and different) password, which you can do by pointing any web browser (on your iPad or computer) to iforgot.apple.com and following the onscreen instructions there.

You Forgot Your iPad Passcode

Given how many passwords we're all required to keep in our heads for various devices, websites, and services, you're bound to forget one now and then. And if you can't remember the passcode you use to unlock your iPad, you're left with a locked device that is of no use to anybody.

That's because every iPad is programmed to lock out all users if the wrong passcode is entered 10 times in a row. It's a safety measure to keep ne'er-do-wells from trying to force their way into a stolen device. Too many wrong passcodes and the iPad is disabled.

(This actually happened to my daughter-in-law, through no fault of her own. Her then six-year-old son—my rambunctious little grandson—got hold of her iPad and tried to break into it by entering random numbers on the Lock screen. Ten tries later and her iPad was completely inaccessible.)

When this happens, the only way to get back into your iPad is to completely erase everything stored on your device, and revert to factory-fresh condition. You can then restore your data if you've backed it up beforehand, that is, and create a new passcode.

The easiest way to erase your iPad is via iTunes. Follow these steps:

1. Connect your iPad to your computer and launch the iTunes software.

2. Wait for iTunes to sync your iPad to your computer and make a backup. Once the sync and backup are completed, click Restore.

3. Your iPad has been reset to factory-fresh condition, and the setup process commences on your device. When you're prompted to restore your iPad, tap Restore from iTunes Backup.

4. Back in the iTunes software on your computer, select your device.

5. Select the most recent data backup and follow the onscreen instructions to restore the backed-up data.

>>>Go Further

APPLE SUPPORT

If you work through the tips in this chapter and still have problems with your iPad, it's time to turn to Apple's official technical support. You can get support at any Apple Store location or online at https://support.apple.com/ipad. There are also many third-party repair companies that specialize in repairing iPads and other local devices; search Google to find one in your area.

Updating and Resetting Your iPad

It's important to keep your iPad up-to-date, which is what updating the operating system is all about. Operating system updates sometimes contain new features and almost always contain all manner of bug fixes. So if you've been having recurring issues with your iPad, chances are they'll be fixed in the next version of the operating system.

Sometimes, however, things can get so gunked up that the only way to fix a recalcitrant iPad is to wipe it clean and reset it to the original factory condition. This should be the method of last resort, however, because it wipes out all the apps and data you've accumulated on your device—although you can easily and quickly restore your data and settings from an iCloud or iTunes backup.

Update to the Latest Version of iPadOS

Your iPad is controlled by an operating system that Apple calls iPadOS. This operating system is updated from time to time, and you want to make sure that your iPad is running the latest iPadOS version.

In most instances, your iPad tells you (via an onscreen notification) that a new version is available and walks you through the update procedure. You can, however, manually check for and install iPadOS updates.

1. From your iPad's Settings screen, tap General in the left column.

2. Tap Software Update.

3. If a software update is available, tap Download and Install and then follow the onscreen instructions.

Factory Reset Your Device to "Like New" Condition

If you've tried everything to get your iPad working again and nothing has worked, you might need to take the drastic step of resetting your iPad to "like new" condition. You should only attempt this device reset when all other steps have failed; it's a big thing.

What's big about it is that resetting your iPad erases all the data and apps stored on your device. You'll lose all your music, videos, pictures, you name it, and have to reinstall everything from scratch.

That said, the "reinstall from scratch" process isn't as dire as it sounds. If you've backed up your iPad in advance (via either iTunes or iCloud), you can reinstall all your backed-up data and settings.

Backing Up Your iPad

Learn more about backing up your iPad via either iCloud or iTunes in Chapter 24, "Managing Files on Your iPad and in the Cloud."

When the reset is done, your iPad is wiped and restarts with the original out-of-the-box setup process. You need to walk through the setup as if you just purchased the iPad, but then you can connect to iTunes or iCloud and reinstall your backed-up files.

It's Not All Good

Resetting Erases Everything

I will repeat this warning. *Resetting your iPad to factory-fresh condition erases all the data, apps, and settings stored on your device.* If you've previously backed up your iPad (to iCloud or iTunes), you can restore your data, apps, and settings. If you have not made a backup, your data and settings will be lost—*permanently*. (You can always reinstall apps you've previously downloaded from Apple's App Store.)

1. From your iPad's Settings screen, tap General in the left column.

2. Scroll to the bottom of the right column and tap Reset.

3. Tap Erase All Content and Settings.

4. Enter your passcode.

5. When prompted, tap Erase. (Or tap Cancel to not proceed.) Your iPad—and all its data, settings, and apps—will be erased.

Restore from an iCloud Backup

When you reset your iPad, everything is wiped clean and your device is back to its factory-fresh condition. After the reset is complete, your iPad's initial setup process starts, and it shows you the "Hello" screen again (discussed in Chapter 1, "Buying and Unboxing Your iPad").

You restore your iPad during this initial setup process—so look sharp at each screen you see and follow these instructions.

1. Proceed through the onscreen setup until you reach the Apps & Data screen, and then tap Restore from iCloud Backup.

2. When prompted, sign in to iCloud with your Apple ID.

3. On the next screen, choose a backup. You'll likely want to choose the most recent backup.

4. The backup starts. When prompted, you need to sign in to the App Store with your Apple ID to restore your apps and purchases.

5. Make sure you stay connected to the Internet during the entire process, which might take up to an hour to complete. When the files are restored, you can finish the initial setup and start using your iPad as normal.

Restore from an iTunes Backup

If you've previously backed up your iPad to your computer, via iTunes, you can restore your device from iTunes during the initial setup process.

1. Proceed through the onscreen setup until you reach the Apps & Data screen, and then connect your iPad to your computer.

2. The iTunes software on your computer should launch. Select your iPad from the available devices.

3. Back on your iPad, on the Apps & Data screen, tap to select Restore from iTunes Backup.

4. In iTunes, select Restore Backup.

5. Select the backup from which you want to restore. (Probably the most recent one.)

6. Click Restore.

7. Keep your iPad connected to your computer until the restore process is complete. When it's done, you can finish your iPad's initial setup and start using it as normal.

Glossary

AirDrop   The feature that enables sharing of media files between an iPad and other Apple devices.

AirPlay   The feature that streams music and videos from an iPad to other Apple devices.

AirPods   Apple's wireless Bluetooth earbuds.

app   A software application running on the iPad or other mobile devices.

App Exposé   The feature that lets you see all open windows for an app.

App Store   Apple's online store that offers apps and games for iPads and iPhones.

App Switcher   The screen mode that displays all running apps in a swipeable carousel.

Apple   The company that created the iPad and iPhone.

Apple Arcade   Apple's subscription game-playing service.

Apple Pay   Apple's mobile payment system.

Apple Pencil   The active stylus designed for use with selected iPad and iPad Pro models.

Apple TV   Apple's app for viewing videos; also the name of Apple's streaming media device.

Apple TV+   Apple's subscription-based streaming video service.

AssistiveTouch   A feature that overlays a group of large icons onscreen for common functions.

Bluetooth   Wireless technology designed to connect two devices.

cloud storage   File storage on Internet-based servers.

Control Center   The panel that swipes in from the top-right corner of the iPad screen and enables control of many common system settings.

Dark Mode   A dark color scheme.

Do Not Disturb mode   An iPad operating mode that hides notifications, alerts, and system sound effects.

Dock   That area at the bottom of every iPad screen that displays the same selected app icons.

earbud   A small earphone that fits inside the wearer's ear, used to listen to music and soundtracks on an iPad or another device.

eBook   An electronic book; a traditional book in electronic form.

email   Electronic mail; messages sent electronically from one computer or device to another.

emoji   Ideaograms and "smileys" used to enhance text messages and emails.

Face ID   Apple's facial recognition system for unlocking selected iPad models.

FaceTime   Apple's video chat service.

FaceTime camera   See *front-facing camera*.

force restart   The process of manually forcing the iPad to restart.

front-facing camera   The camera above the iPad display, typically used for selfies and video chats.

HDR High dynamic range, a process that combines the results of multiple photos to create a single high-contrast photo.

Home button The small round button beneath the iPad screen when held vertically. Not present on iPad Pro models.

Home screen Those screens that host the icons for all the apps installed on an iPad.

HomeKit Apple's smart home technology and system.

hotspot A location where you can get Internet access via Wi-Fi.

iCloud Apple's cloud-based storage service.

identity theft The fraudulent acquisition and use of an individual's personal information.

in-app purchase The purchase of additional services from within an app or game.

iOS Prior to iPadOS, the operating system used by the iPad. Still used on Apple's other mobile devices.

iPad Apple's popular tablet computer.

iPad Air A more full-featured version of the basic iPad.

iPad mini Apple's smallest and most affordable iPad, with a 7.9-inch screen.

iPad Pro The iPad designed for business productivity, with 11- and 12.9-inch models.

iPadOS The operating system used to control the iPad.

iPhone Apple's smartphone.

iSight camera The rear-facing camera on the back of the Apple iPad.

iTunes Apple's software that enables users to manage their iPads and iPhones on their Mac or Windows computers.

iTunes Store Apple's online store that offers music and videos for purchase and rental.

Keynote Apple's presentation app for the iPad and iPhone.

Lightning connector   The port on the bottom of most iPads that is used to connect the iPad to the power adapter, computers, and other USB devices.

Lock screen   The screen that appears before you unlock your iPad.

Magic Keyboard   Apple's wireless keyboard/trackpad, designed to convert an iPad to computer-like use.

malware   Short for malicious software—viruses, spyware, and other files that can damage a computer or mobile device.

Memoji   An emoji-like avatar that tracks your facial movements.

mouse   A handheld input device that is dragged across a flat surface to move a cursor across the screen.

multitasking   The capability of using more than one app at the same time.

Multi-Touch Display   Apple's touchscreen display used on the iPad.

newsreader   An app that consolidates news items from multiple sources.

Night Shift   The feature that enables you to adjust the color temperature of the iPad display for night-time viewing.

Notification Center   The iPad screen that displays system notifications and app alerts.

Numbers   Apple's spreadsheet app for the iPad and iPhone.

On/Off button   The physical button at the top-right corner of the iPad (when held vertically) used to put the device to sleep and wake it up.

operating system   A core software program that controls a device's underlying hardware and operations.

Pages   Apple's word processing app for the iPad and iPhone.

passcode   On the iPad, a six-number code that must be entered to unlock the device.

Perspective Zoom   A type of parallax effect that displays a slight motion on the Home or Lock screens when you tilt your iPad.

pixel   A single dot or picture element that makes up a photo or video picture.

post   See *status update.*

predictive keyboard   A virtual keyboard that attempts to figure out what you're typing and enter that word for you.

rear-facing camera   See *iSight camera.*

reset   Wiping your iPad of all data, settings, and apps and returning it to factory-new condition.

Retina display   The iPad display that offers 300 pixels or more resolution.

Safari   Apple's web browser app.

Safari Reader   A special reading mode, within the Safari browser, that makes some web pages easier to read by removing ads, images, videos, and other extraneous elements.

Scribble   Functionality that enables handwriting in online forms and text fields with an Apple Pencil.

selfie   A picture you take of yourself.

Siri   The voice-activated software that functions as a virtual personal assistant on your iPad and other Apple devices.

Slide Over   A feature that enables you to open a second app onscreen without closing the one you're currently in.

Smart Connector   The wireless connector unique to the iPad Pro, designed to connect the Magic Keyboard, Smart Keyboard, and Apple Pencil.

smart cover   An iPad cover that wakes the unit from sleep when open and puts it in sleep mode when closed.

Smart Keyboard   Apple's wireless keyboard designed for use with the iPad Pro.

social media   See *social network.*

social network   An Internet-based service that hosts a community of users and makes it easy for those users to communicate with one another.

spam   Also known as junk mail, any unsolicited email advertisement.

Split View   The mode that makes two apps active onscreen at the same time.

status update   A short message (with text and/or images and video) that updates friends on what a user is doing or thinking.

streaming media   Music and video that are transmitted in real time to a connected device.

tablet   See *tablet computer*.

tablet computer   A small computer in the shape of a handheld tablet. The iPad is a tablet computer.

text messaging   Short electronic messages sent from one device to another.

touch gestures   Taps and motions on the iPad screen that perform various common operations.

Touch ID   Apple's fingerprint-sensing system for unlocking selected iPad models.

touchpad   An input device in the form of a small touch-sensitive panel, typically incorporated into notebook computer keyboards.

touchscreen   A device display that can be operated by touch gestures.

URL   Uniform Resource Locator, the address of a web page.

USB   Universal Serial Port, a type of connection common to personal computers.

USB Type-A—The standard USB connector used on most computers and peripherals. Slightly larger than the USB Type-C connector.

USB Type-C—The type of USB port on the bottom of iPad Pro models that is used to connect the iPad to the power adapter, computers, and other USB devices.

video chat   A face-to-face onscreen chat between two users.

VoiceOver   A feature that describes out loud any screen element or text on the iPad.

wallpaper   The background image on the iPad's Home and Lock screens.

Wi-Fi   Short for wireless fidelity, the wireless networking standard used by most computers and connected devices today.

widget   A small app designed to perform a single specific function.

YouTube   The Internet's largest video-sharing community.

Index

Answers to Your Technology Questions

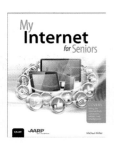

The My...For Seniors Series is a collection of how-to guide books from AARP and Que that respect your smarts without assuming you are a techie.
Each book in the series features:

- Large, full-color photos
- Step-by-step instructions
- Helpful tips and tricks
- Troubleshooting help

For more information about these titles,
and for more specialized titles, visit
informit.com/que